Table of Contents

WARNING:

Technical information contained in this manual concerning the modification of firearms (as well as their use and adjustment), and all information concerning the use of ammunition is necessarily a reflection of the author's beliefs based on past experience. As it is unlikely that the reader will be able to precisely duplicate efforts made by the author, the information contained herein should be used as a general guideline and approached with great care and caution. Neither the author or the publisher assumes any responsibility for the use or misuse of the contents of this manual.

Foreword

I first heard of Dan Boone in the middle of a Sunday-morning sermon. When I heard the name, I knew this guy had to be "into" guns. How could a guy with the name Dan Boone not be?

Sure enough, the pastor continued, telling us he had a rifle that didn't hit the mark– similar to how most of us miss the mark when it comes to our daily lives. The rifle was getting straightened out with new sights mounted by a skilled gunsmith; the pastor picked the best man he knew for the job, and that man was Boone.

A week after hearing about Boone and his work on the preacher's rifle, I had the pleasure of meeting the gunsmith in the flesh. I discovered him to be both a gentleman and a scholar, working to secure a double degree in physics and psychology, creating a new branch of psychology in the process with the research he was engaged in.

He also had a manuscript along for me to check out. Fortunately it was about firearms rather than psychology or physics, something my own school records suggest I'm marginal at when everyone is being generous.

Boone's manuscript was a pleasant surprise and

enjoyable reading. I made a few suggestions to him–– mostly having to do with layout of the manuscript rather than the details in it; it was obvious as I looked through it that he knew his subject and there was little to correct in that area. A few weeks later he had the whole book expanded and rewritten, ready to submit to his publisher.

Boone's manual is an important document to have in the hands of American shooters since the gun grabbers in our country are hell bent to disarm honest citizens. Boone's book helps counter this trend by showing readers how to arm themselves with a quality— but inexpensive–firearm suitable for target shooting, hunting, or self defense.

This book also reveals the secrets of reviving the old "war horses" that many shooters have laid aside for newer rifles. These dusty bolt-action firearms can often be purchased for a song; with Boone's know-how, it's possible to transform such a rifle into a tack driver with an interesting historic pedigree.

I'm hoping all the readers of this book enjoy it as much as I have.

Duncan Long

Introduction

Highly accurate rifles are commercially available in the United States at the present time, but they typically come at a very high price (frequently well over a thousand dollars). It is my belief that most people in the U.S. do not have a couple of thousand dollars to lay out for a highly accurate rifle system, but many would like to have one. This manuscript describes how such a highly accurate weapon might be produced at nominal cost.

This manual will provide instructions on how to modify inexpensive, military surplus bolt action firearms, producing a rifle with accuracy capabilities ranging from greatly improved to at or near sniper capability, and to do so at a fraction of the cost of a commercial rig. Obviously, hunters and sportsmen who enjoy target shooting could benefit from this book as well, and it would probably benefit all of us to do this while we still have the legal right.

This book is necessarily written as though the reader knows very little about firearms. Although some readers may be highly knowledgeable about firearms and may consider some of the material described here to be too simplistic, it is the author's desire that anyone who wants a sniper-grade rifle be given the opportunity to obtain and/or build one, including those that are only minimally knowledgeable about guns. Every attempt will be made to make the information complete, detailed, informative, and as simple as possible. It is my hope that those who are highly knowledgeable about firearms will be more acceptant of the style of presentation, given the reason for it.

One thing needs to be understood right from the beginning: the reader is going to necessarily be involved in various minor aspects of gunsmithing. This assumes that the reader either possesses some skill with basic tools or can learn the skills that he needs. None of the techniques used are difficult or complicated, but if you are one who has virtually no mechanical abilities, it might be prudent to get a friend to help and advise you in your work.

What this book is designed to do is to help the reader select an inexpensive military surplus rifle which is suitable for conversion to sniper, then to take the reader through the process of making the sniper conversion step-by-step. At this point one might ask "Why a military rifle? Why not a commercial rifle?" This is a valid question and one that deserves being addressed. I have always had a preferential leaning toward military surplus rifles over commercial rifles for several reasons. First, military rifles are typically very rugged rifles. There are few nations that would be so unwise as to arm their troops with poorly made rifles, assuming they expect to win the conflict they are arming for. Thus, military rifles are made (in my opinion) much stronger and more reliable than most commercial rifles. This is not to say that all commercially made rifles are necessarily weak and easily damaged, or malfunction frequently. Indeed, some commercial rifles have had such a fine track record for accuracy and dependability such that they have been adopted by both military units and police SWAT teams as sniper rifles. However, generally speaking, commercial rifles are intended to function as a hunting tool, to be used intermittently across a month or so each year during hunting season. They are not designed to undergo the rigors of combat on a daily basis as military rifles are. As a result, I believe that if one were to compare military rifles against commercial rifles with equal amounts of use and abuse on each, you would find far more breakdowns and part failures in many commercial models than in military models.

Second, military rifles are designed to function under all kinds of adverse conditions, in rain or snow, mud or sand, or what have you. Commercially made rifles are designed to function well in relatively clean environments. It is assumed by the manufacturer that the owner of a commercially made hunting rifle is going to keep the firearm clean. Most gunsmiths will tell you that the most common cause of commercial firearm failure is dirt. Commercial manufacturers assume that the average hunter is not going to be crawling through a muddy field with his rifle cradled in his arms. Accordingly, when faced with truly dirty, adverse conditions, the commercial rifle often does not function properly. Semi-autos are especially sensitive to dirt in the action. Commercial rifles are just not designed to handle the kind of crud that comes with military/sniper use.

Last of all, one must consider cost. Because military rifles are surplus, they can frequently be purchased at very reasonable prices. Cost tends to vary according to condition, availability and demand, but it is generally possible to obtain a good military surplus rifle for 1/5th to 1/10th the price of a good commercial rig, or even less if we are comparing the surplus supply to top-of-the-line commercial rifles (i.e. with sniper-grade accuracy).

Highly accurate (and expensive) commercial rifles are expensive because someone has already done all or most of the accurizing described in this book before the rifle was shipped to the wholesaler. The main objective of this book is to demonstrate the process of hand-building a sniper rifle. It is not a manuscript on how to make a rifle look pretty or attractive. Subjects such as bluing, plating, stock carving, checkering, etc., are left to other books, of which there are many on the market. The objective of this publication is not aesthetics but function, and is best embodied by the old adage "beauty is as beauty does". It will be left to the reader to take whatever steps he deems necessary to make the end product more attractive, since "beauty" is largely a matter of individual taste anyhow.

Chapter 1

The Sniper Rifle

Most military rifles have a special production number designated for sniper models. For example, the number 4 mark 2 British SMLE rifles were used for the common troop usage, but their sniper rifle was designated the number 4 mark 2 (T), the "T" indicating a tang mounted sight. This particular model also came without the charging slot for use with stripper clips (since this was not needed for sniper applications) and was drilled and tapped for telescopic sight mounts. Sniper models typically have minor modifications which makes them differ slightly in appearance from the standard rifle model. In the American Springfield rifle, the sniper designation was the 1903-A4, which had a slightly different stock from the regular 1903-A3, the shape of the bolt handle was a little different (to allow for clearance when a telescopic sight was installed) and it was issued without iron sights, since it was assumed that iron sights would not be used in a sniper application. The point is, military rifle models which are desig-

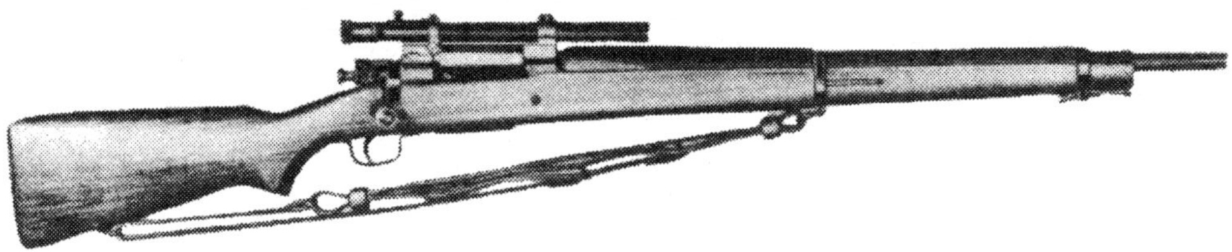

The American Springfield Rifle as manufactured in the sniper configuration and designated U. S. Rifle, Cal. .30, M1903A4 (Snipers) With Telescopic Sight M73b1 (Weaver No. 330C) and Sling M1907.

nated "sniper" models will usually be marked differently and will look slightly different without the telescopic sight mounted. Of course, with the telescopic sight mounted, it takes on a considerably different look. When one obtains a military surplus rifle, it may or may not be a sniper model. This doesn't necessarily have any great significance except for the fact that sniper models are more desirable as a collector's item, even if it is not complete (sniper models are not complete unless they are fitted with their military issue telescopic sight). The use of modern accurizing methods (e.g. glass bedding) which were not known or used at the time of most surplus rifles' manufacture, allows even the more common military rifle models to perform with a very high degree of accuracy.

Military surplus sniper rifles are exorbitantly expensive on the open market, commonly going for one to two thousand dollars (or more) wholesale. We are not talking about modern, state-of-the-art rifles here, but military surplus often going back to World War II and before. There are several reasons for the high prices. First, they are a high priority item for military collectors, who are willing to pay high prices to get one. Second, there were much smaller numbers of sniper rifles made by the various governments, and rarity usually translates into higher prices. Third, if the sniper rifle was issued by some governmental agency, it was hand-tuned for accuracy. The process of fine hand adjustment to get all of the accuracy out of the rifle that it was capable of producing, along with the added expense of the telescopic sight system used on the sniper rifle, made it a costly item.

Sniper rifles were required to be highly accurate

over very long distances. I have seen general statements made in recent advertisements to the effect that sniper rifles are accurate for distances of "...up to a mile away!" This is certainly an exaggeration. Some of the large frame bolt action rifles which have come out in the last 10 years or so chambered for the .50 caliber machine gun cartridge are capable of telling accuracy at that distance, but this is certainly the exception to the rule. The common military cartridges will not shoot accurately at such a range, but they can be accurate, at least in theory, up to 1000 yards on a target the size of a man. During military service, I qualified with an M-1 Garand at 1000 yards, but the targets employed at this distance are large enough to be used as a door for an elephant pen. We need to realize that theory is one thing and practice is quite another. It would not only take an exceptional rifle equipped with an exceptional sight system to produce accurate shots on a man-sized target at 1000 yards, it would also require an exceptional marksman. The Shepherd rangefinding telescopic sight claims one minute of angle (MOA) accuracy out to 1000 yards for certain cartridges such as the 7.62 NATO, the 7.62 x 54 mm rimmed Russian, the U.S. M-1 30.06 military ammo, and .300 Win. magnum (See Shotgun News,June 20, 1994, p. 55. More about the Shepherd scope in section V of the manual). If the rifle (equipped with the Shepherd scope) itself is capable of one MOA accuracy out to 1000 yards, assuming that one could hold the rifle steady enough to keep it on target, this translates into all shots going approximately into a 10 inch circle at that distance. In his description of the practical rifle course at Jeff Cooper's gunsite ranch in Paulden, Arizona., Bruce Clayton (*Thinking About Survival*, 1984, Paladin press, Boulder Colorado) states that scopes were zeroed at 200 meters. This would supposedly place all shots in the kill zone at any range up to 300 meters. Firing was done at ranges of 200, 300 and 400 meters. Clayton's statement was that out to 300 meters, just about everyone grouped their shots pretty well, but at 400 meters people had trouble "finding the paper [of the target]". However, he also states "In all fairness, though, I should point out that some students were making disgustingly tight groups at 400 meters (p.87)." Perhaps the 400 meter mark is the point at which the rifle's innate accuracy no longer compensates for a lack of skill on the part of the shooter. There may be a way to get around the 300 meter barrier, however, and this will be discussed later in this manual.

It is interesting that many WWI and WWII military rifles have open v-notch sights calibrated out to 2000 meters or farther. Anyone who has ever set these sights to their maximum distance will know that sighting in on a target in this way results in the rifle being elevated like an artillery piece. It makes for interesting conversation, but I seriously doubt that anyone was ever hit at these distances by some marksman using such crude sights. In fact, a range of a thousand yards is pushing the limits of any sniper's ability. Most shots are at closer distances, probably 300 to 700 yards.

When I was in Vietnam, the Marine Corps was using the model 700 Remington as a sniper rifle. It was equipped with a heavy target barrel and a Redfield 3x9 variable scope. Unertl target scopes have also frequently been used in the past, and the most recent USMC Sniping manual (FMFM1-3B) calls for a 10x Unertl scope.

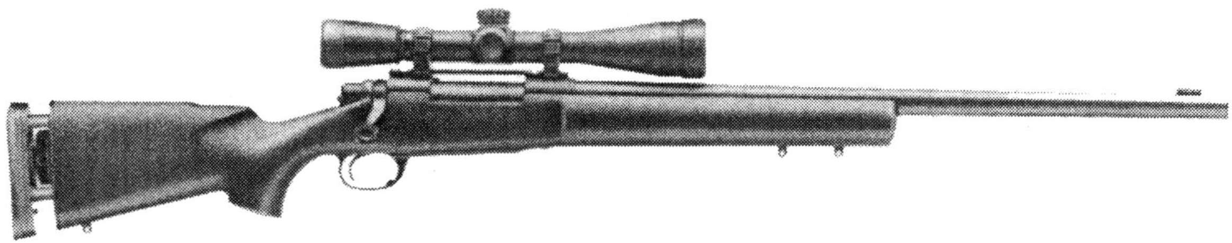

The U. S. Army M24 sniper rifle is built on a Remington 700 action and fitted with an adjustable length stock. It features a high quality 10X scope that performs well under low light conditions.

Chapter 2

Selecting the Rifle to Convert

Many people already own a military surplus rifle that they may want to convert. Some rifles, however, are not suitable for conversion. In the paragraphs to follow, the author will describe military surplus rifles which are (in the author's opinion) not suitable for conversion to sniper configuration. If the reader already owns one of these unsuitable rifles and wants to try his hand at converting it anyhow, he is welcome to "give it a go," but should not expect much in the way of improved performance. On the other hand, with some of these unsuitable rifles, going through the accurizing process can't hurt them and might actually improve their functioning. It is unlikely, however, that they would ever generate a level of performance anywhere near what you would expect for a sniper rifle. The only real risk in doing such a conversion, however, is that the time and money expended on the rifle is not really worth what you get at the end of the accurizing process. Some might feel that it is worth both time and money expended for the experience alone, especially if he has a more expensive rifle that he wants to convert, but would like to gain some practical hands-on experience before going for the "gold ring."

We will make an assumption at this point that the reader does not yet have the rifle that he will want to convert, but that he will select it and buy it with the idea of making a sniper conversion. With this in mind, there are several things which need to be considered:

The first important consideration is the grade of barrel in the rifle. A match grade barrel is preferable but not generally found in military surplus rifles. However, certain rifles have extremely high grade barrels which approach match grade, and some are even equivalent to match grade. Others fall appallingly short of this standard.

A second consideration is barrel length. In the old days it was believed that the longer the barrel, the higher the muzzle velocity and the more accurate the rifle. This is why many of the old Plains rifles and Kentucky rifles had barrels of 3 to 4 feet in length. Modern experimentation has shown that muzzle velocity does increase dramatically as barrel length increases, but only out to about 26 inches or so (this varies to some extent with the rifle). At longer lengths, muzzle velocity does increase, but in increments too small to be practically useful. Increases in barrel length beyond 26 inches simply add additional poundage which must be carried in the field (a relatively undesirable characteristic) without adding enough advantage in increased muzzle velocity to make the extra weight worthwhile. Longer barrels do provide more accuracy in the sense that they provide a longer sight radius if iron sights are being used, but if a telescopic sight is being employed, this advantage is lost (because telescopic sights do not depend on the length of the barrel for increased sight radii to provide accuracy, but on their own optics). Since telescopic sights are inherently far more accurate than iron sights, and with the use of telescopic sights a longer sight radius is not necessary, it is easy to see that the ideal barrel for a sniper rifle would be around 25 to 26 inches (give or take a few inches). If the barrel is an inch or two longer than this, the increase in weight will probably not be enough to be substantially noticeable in the field. If it is significantly longer than 26 inches (and some are. The old time practice of using very long barrels was carried well into the first half of 20th century.), it can always be shortened. In fact, under certain circumstances it is advisable to shorten the barrel. The procedure will be described later in this book. If one has a rifle with a good barrel which is very long (say, 31 inches, a common length in some rifles), and if he does not want to shorten the barrel, this is fine also. How much weight that one is willing to carry is an individual choice, and the longer barrel certainly won't hurt the accuracy of the firearm.

A barrel that is too short is quite another matter, however. If the barrel on a surplus rifle is substantially shorter than 25 to 26 inches, it may not be possible to attain high enough muzzle velocities to provide accuracy over very long distances. Rifles with carbine (short) barrels are often much handier to carry in the field, and can be very accurate at shorter distances, but should be restricted to much shorter ranges, optimally around 150 yards and maxing out

When it comes to stopping power the U. S. M14 (top) was far superior to the M16 (bottom) with its smaller caliber.

at around 300 yards For this reason, short barreled military rifles should be avoided as a source for possible sniper conversions.

A third consideration of importance is the caliber of the barrel and the specific cartridge the rifle is made for. When we talk about caliber, we are talking about bullet diameters/barrel groove diameters rather than specific cartridges, although it is a common practice to use the terms "caliber" and "cartridge" interchangeably.

Both very small and very large calibers should be avoided. Very small calibers (e.g. .22 calibers) are too light to carry long distances with accuracy, and do not have a great deal of stopping power. The next statement that I am going to make will no doubt raise the ire of some, but it needs to be said. I believe that one of the biggest mistakes the U.S. military ever made was the replacement of the M-14 rifle with the M-16. The M-14's 7.62 NATO (.308 Winchester) cartridge is a very efficient round and a good man-stopper, and replacing it with a rifle chambered for the .223 Remington is a serious step backward. It would seem that the military's objective of "kill the enemy" has been changed to "wound the enemy", because this is exactly what the .223 cartridge is best suited for when it comes to large North American animals (man included). For those that disagree, I would recommend that you take notice that the .223 is illegal in most states for use in the hunting of deer-sized game (and larger) because of its inability to either stop game in its tracks or to provide a clean kill. If it is no good as a deer-stopper, it will certainly not be any good as a man-stopper, and what is worse, a wounded soldier can (and will) return fire.

Very large calibers (e.g. the .45) should also be avoided. These utilize bullets so large and heavy that their trajectory resembles a rainbow. Firing a very large caliber rifle at a target at considerable distance would require elevating the rifle like an artillery piece. Large heavy bullets also generate considerable amounts of recoil, a problem which is best avoided if the rifleman is expected to practice with his rifle to any great degree.

Medium range calibers in bullet weights of about 140 to 180 grains (depending on caliber) are best. Medium range calibers would be 7mm to 8mm in European made rifles. Some 6.5 mm military rifles have great accuracy at long distances, but I personally feel that the best performance is possible with slightly larger calibers. This is a personal preference, however, and the more accurate 6.5 mm military rifles will be mentioned later, in case the reader has a preference for one of these. In American calibers, typically .30 to .32 caliber is best. Virtually all American made military rifles manufactured for U.S. use, from the 30-40 Krag to the U.S. M-14 rifle, are .30 caliber. British Enfield rifles with calibers designated .303 British are actually 7.7 mm calibers.

Ammunition is also an important consideration. It is advisable for one to avoid rifles which are chambered for rare, esoteric, or obsolete cartridges. Common cartridges such as 30.06, 7.62 NATO (.308 Winchester) or 8 mm Mauser are so readily available and in common usage that virtually all types of ammunition are available, from cheap military surplus (for practice) to heavier soft-tip hunting ammunition, to high grade match ammunition. With this in mind, one can see that although the 8 mm caliber is generally acceptable for sniper conversions, a rifle such as the Steyr-Mannlicher M95 which shoots the rarely seen 8 x 56 mm cartridge would not.

It should be noted that any military rifle can typically be rebarreled or the existing barrel can sometimes be rechambered to accept a different cartridge. For example, the American pattern 17 Enfield in 30.06 can be rechambered to accept a .300 Winchester magnum. However, rebarreling and/or rechambering is work that will require a properly equipped gunsmith. This type of work does not usually come cheap either. For this reason, this manual avoids situations (and rifles) which may call for rebarreling or rechambering, and emphasizes the use of military surplus rifles in their original military configuration, so the reader can do all of the work needed himself.

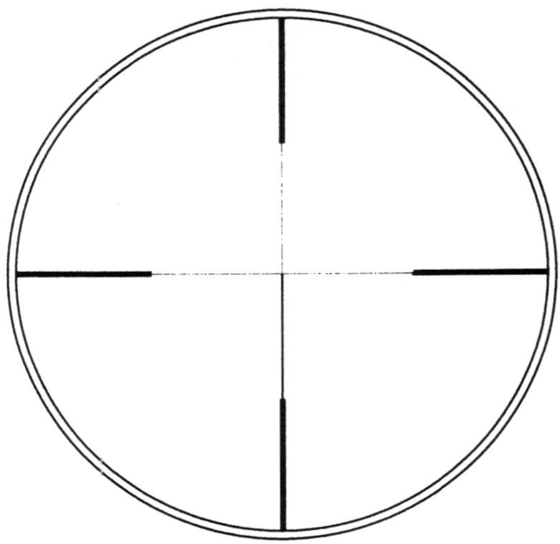

Chapter 3

Rifles to Avoid and to Select

As mentioned earlier, some military rifles are not suitable for conversions. Although the general principles of what would constitute an unsuitable firearm for conversion has been given earlier, the following is a partial list of specific rifles which are not suitable for conversion along with an explanation of why they are not.

1. Any antique military rifle designed for use with black powder. The modern smokeless powder used in sniper rifles generates much higher pressures than that produced by black powder, pressures which would blow the black powder rifle apart. Also, most of these rifles are single shot weapons, a situation which is decidedly disadvantageous for a sniper. Examples of these types of rifles would include 1873 trapdoor Springfields, British Martini-type rifles, rolling block Remingtons and (with apologies to Matthew Quigley) the old Sharps rifles.

2. Italian Carcano rifles, both models 1891 and 38: Inappropriate caliber, barrels usually too short, crudely built actions. Some of these were converted from 6.5 Carcano to 8 mm Mauser during WW II; dangerous to use.

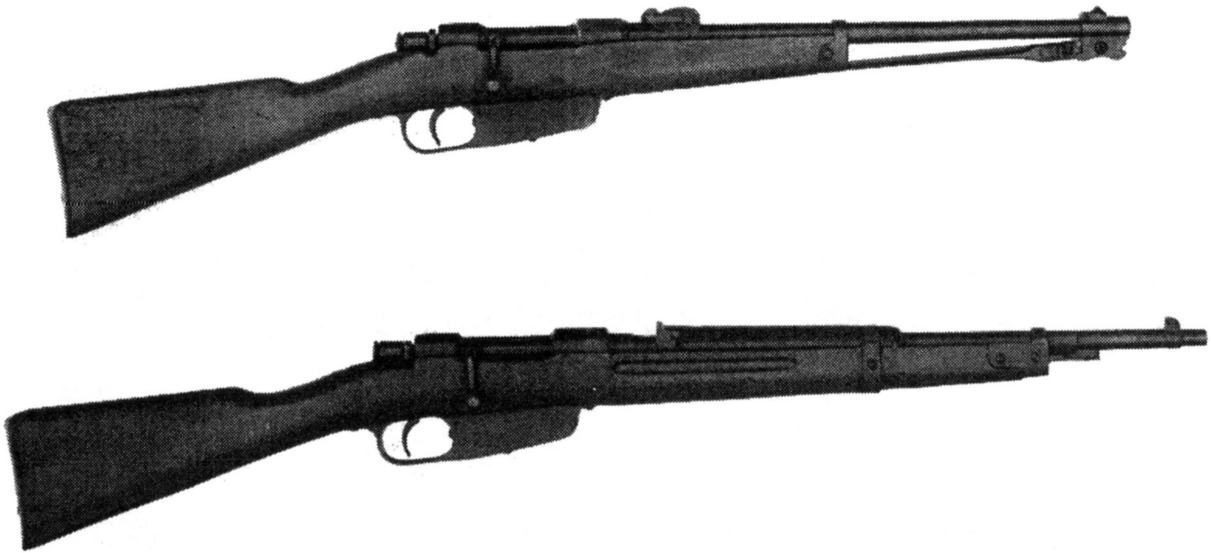

The Manlicher-Carcanos, 1891 (top), and 1938 (bottom) are of inappropriate caliber, have barrels that are usually too short, and have crudely built actions which makes them unacceptable for conversion to an efficient sniper rifle. One has to wonder what really happened when the Carcanos, 1938 (bottom) is the type used to assassinate President John F. Kennedy.

3. Austrian Steyr-Mannlicher, model 95: obsolete ammunition, typically short-barreled as found on the surplus market.

4. Type 54 Chinese rifles or type 44 Russian rifles: These are carbines and barrels are too short for use. The cartridge it uses may pose a problem as well. The 7.62 x 54 mm rimmed is an excellent cartridge, but it is a Russian/Communist Bloc nation round. It is not really rare, but at least in the United States it is not real common either. At the present time, no American manufacturer makes ammo for this rifle. All existing supplies to the United States come from imports which have been blocked by President Clinton at the time of this writing. There is no way to tell whether the public will be able to resupply themselves with this cartridge once existing supplies run out.

5. French 1936 MAS: 8 mm Lebel ammo is not generally available, the bolt action is ergonomically poorly designed, accuracy in general is not good.

6. French 1886 Lebel: inappropriate ammunition, barrels are innately not very accurate.

7. Model 93 and 95 Mausers: These are strong rifles with good barrels of adequate caliber, but they were manufactured without gas vents machined into the bolts. As a safety feature, it is recommended that these rifles not be fired, since the lack of gas vents virtually guarantees that hot gases from a ruptured primer will come right back into the face of the shooter. There are much safer designs available, so just hang these on the wall and leave them there.

8. Mannlicher-Berthier: obsolete ammunition.

oooooo

Some military rifles are probably inappropriate for conversion because their calibers are a little too small, but because of their fine accuracy the reader might want to use them anyhow. These would include the following:

1. The Japanese Arisaka type 38 in 6.5 mm Japanese cartridge, a cartridge that is sometimes hard to find in the U.S. For many years this WW II rifle was seen as Japanese junk. It is now being recognized that in many ways it was a fine military rifle, but actions are rarely sporterized because the cost of such conversions far exceeds the worth of the rifle itself. The rifle is generally worthwhile only if it is utilized in its original military configuration. Its capacity for fine accuracy, even using relatively crude sights, is excellent. It is not at all unusual for a model 38 to shoot 100 Yards. Groups under 2 inches (bench rest) without any modification. The actions are also incredibly strong. The author had a friend, a Colorado gunsmith by the name of Don Grimes, who told of a customer that had complained of excessive recoil in a Japanese rifle he had purchased. When Don checked the rifle out, he discovered the man had purchased a type 38 in 6.5 mm caliber, but was firing 7.7 mm ammunition designed for the type 99 Japanese rifle. The 7.7 mm ammo shouldn't have even chambered in the model 38, but it did, and by all accounts it should have blown up with the first round fired, but instead was drawing out the 7.7 mm bullet to 6.5 mm diameter before exiting the muzzle. This might seem a wild story to some, if it were not for the fact that a similar story surfaced in the May, 1959 edition of American Rifleman magazine. In this story, a man had purchased a type 38 Arisaka and chambered it for 30.06 without reboring the barrel. He took it to a gunsmith because he couldn't figure out why the recoil was so bad. The gunsmith couldn't believe what he was seeing; he fired several rounds of 30.06 ammo through it (no doubt from a distance with a long string attached to the trigger) with no effect. As

before, upon firing, the .30 caliber bullet was drawn out to 6.5 mm. He sent the rifle to the NRA for examination where more rounds were fired through it, astounding many more with the strength of the action. Stories of torture tests on Arisaka actions abound, and there is no doubt that it is exceptionally strong as well as unusually accurate.

The type 99 Arisaka was manufactured later than the type 38 (during WW II) when Japan decided (wisely in my opinion) a larger caliber was needed for their war effort. Because of the rush to manufacture in meeting wartime demands, model 99 rifles are much lower in quality than type 38s, and not nearly as accurate.

Incidentally, if you have an Arisaka with an original dust cover, you might want to remove the dust cover before putting the rifle to any serious use. The dust cover was a piece of semi-circular sheet metal which slid into grooves machined into the sides of the Arisaka action. The dust cover completely covered the action to protect it from dust and anything else that might damage it. When the bolt handle was lifted, it engaged in a slot cut into the dust cover, allowing the cover and bolt to be pulled to the rear together, permitting the ejection of a spent cartridge. When the bolt was closed, the dust cover automatically returned to its place, covering and protecting the action. It was one of those ideas that looks good on paper, but doesn't prove out very well in the field. They generated so much noise that even the Japanese soldiers removed them and threw them away. This is why Japanese rifles with their original dust covers are so hard to find anymore. If you have one, I would keep the dust cover, because the rifle is worth more with it than without it, but I would remove it before going into the field. The racket that it makes could easily give your position away.

2. The model 1896 Swedish Mauser in 6.5 x 55 mm cartridge. This is another rifle with legendary accuracy. Sights are crude and the rifle still routinely turns in 100 yard groups of 1.5 to 1.75 inches (bench rest). Also, since the Swedes never used corrosive ammunition, many surplus rifles purchased have immaculate bores. Ammunition for this rifle is occasionally difficult to find.

3. Moisin Nagant rifles, including the Russian model 91 and Finnish models 91, 27, 28, 28/30. All are chambered for the 7.62 x 54 mm rimmed cartridge, a Russian cartridge whose fate in the U.S. is in question at the present time due to an import ban imposed by President Clinton. The cartridge itself is a good one, equal approximately to the .308 Winchester, but it may become hard to find, especially in match grade. The Russian rifles are accurate, but these cannot begin to compare with the accuracy of the Finnish models, especially those manufactured by Sako with heavy barrels. These barrels could easily be considered match grade. Their innate accuracy potential usually exceeds the ability of the shooter to hold the rifle steady. Incidentally, some early Moisin Nagants were converted to 30.06; these are not safe to fire.

Rifles Recommended for Sniper Conversions (in order of preference)

1. Springfield rifles in 30.06 cartridge, to include models 1903-A1 and 1903-A3. The model 1903 (without the "A" designation) is also a good choice, but only in receivers manufactured at Springfield Armory with serial numbers over 800,000 or receivers manufactured at Rock Island arsenal with serial numbers over 285,507. Receivers

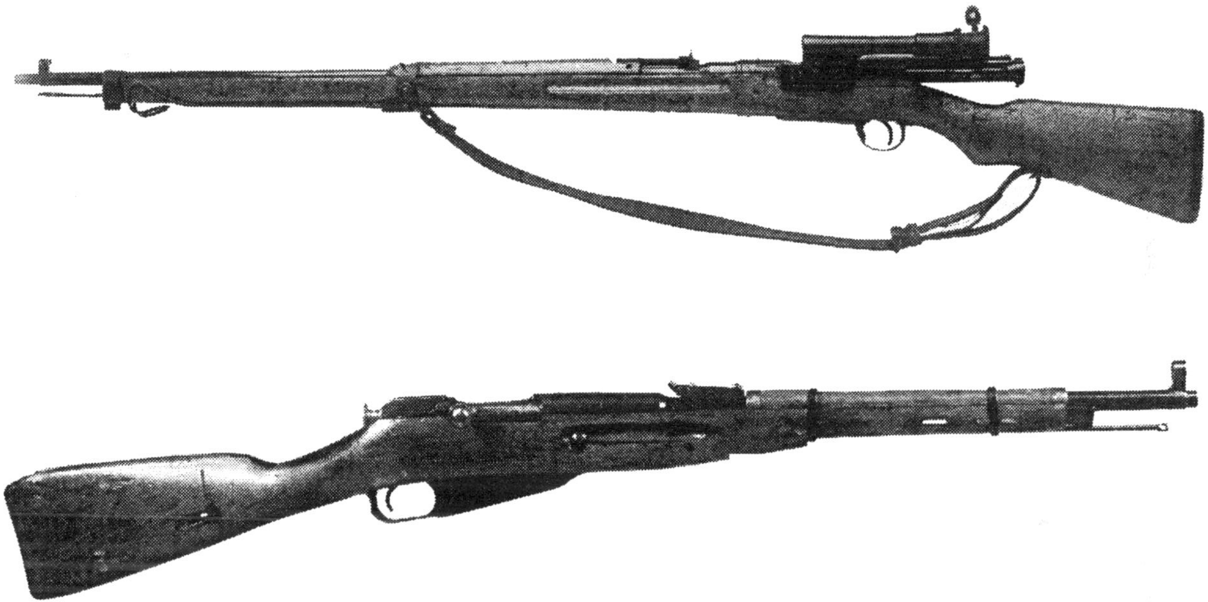

Two mid-range rifles that have marginal calibers but afford superior accuracy are the type 99 Arisaka (top) and the Nagant Carbine M91/30 (bottom).

with serial numbers lower than those given above were heat treated incorrectly and are thus liable to explode on firing. All those with the "A" designation are safe.

2. The pattern 17 American Enfield in 30.06 cartridge.

3. Mauser rifles in 7 mm to 8 mm calibers. Generally, any model 1898 rifle or any other Mauser based on the 1898 action is suitable. These are all very strong and are chambered for cartridges which are widely available anywhere in the U. S. (and the rest of the world as well). Accuracy of barrels vary from model to model. Israeli K98k rifles, Venezuelan FN made 7 mm target rifles and the Brazilian model 1908 in 7 mm are all highly accurate.

4. The pattern 14 British Enfield in .303 British. The pattern 14 and the pattern 17 are identical except for the cartridges they are chambered for. These are highly accurate and very strong, being preferred receivers for magnum conversions in sporters. The commercially made Remington model 30, 30S and 720A are the same as the pattern 17 Enfield except with different calibers, and changes to the rifle's appearance (e.g. it has the Enfield's protective "ears" at the rear of the receiver machined off, a different set of sights and a more aesthetically pleasing stock). Functionally, these are essentially the same.

5. The various models of the short magazine Lee-Enfield (SMLE) in .303 British. These are found with barrels which range from fairly accurate to highly accurate. All could benefit from the accurizing procedures described in this document.

SMLEs often present what would be, for most other rifles, excessive barrel tolerances, but the British found a way around the problem (more about this later). Overall, the SMLE is an excellent battle rifle and well worth the sniper conversion.

At some point the reader is going to notice that all of the

rifles I have dealt with so far are bolt action rifles. There is a good reason for this. It is my personal opinion that the best action for a sniper rifle is the bolt action. Some modern made sniper rifles utilize semi-automatic actions, such as the Russian Dragunov, a sniper rifle designed with a barrel chambered for the 7.62 x 54 mm rimmed cartridge, on a frame which closely resembles that of the AK-47. There are other semi-autos in use as sniper rifles as well. It is still my opinion that the finest accuracy is possible with the bolt action rifle. The fact that the U. S. Marine Corps is still using bolt actions for their sniper rifles is an indication that I am not alone in my belief and preference. The bolt action rifle will probably continue in use as the weapon of choice for sniper configurations, since the primary concern in situations requiring a sniper is fine accuracy over long distances rather than rapidity of fire. In fact, in my opinion a man armed with a good, accurized, military bolt action is not at any tactical disadvantage when compared to a modern soldier armed with one of the .22 caliber "wonders", and is in fact at a distinct advantage, being better armed with a longer reaching caliber having much better stopping power.

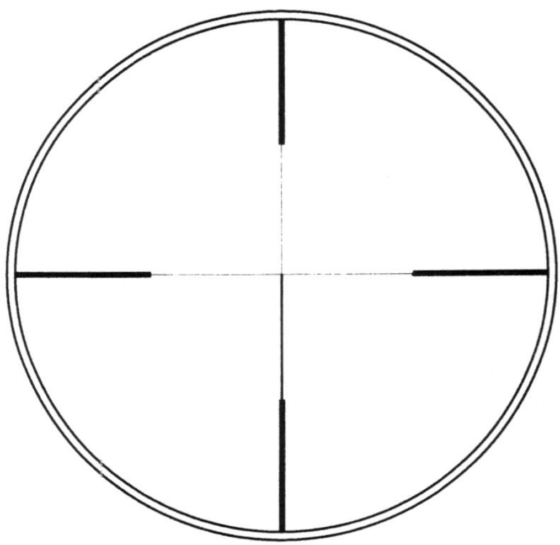

Chapter 4

How to Find a Conversion Rifle

Once the reader has decided what rifle he is interested in, the time has come to lay out the cash and make the purchase. This can be done in two ways:

1. You can begin mulling your way through gun shops and pawnshops looking for the rifle of your choice; when you locate the rifle of your choice, you can buy it at retail prices.

2. Or, you can locate a federal firearm licensee, tell him you want to order a rifle through him, and ask to look through gun advertisements which he may have, such as Shotgun News or Gun List. This latter method can save you both leg-work and money. You can generally negotiate a price for the rifle, and cost plus ten percent is a relatively common deal, with the customer paying the shipping costs. This will allow you to realize a substantial savings over any purchase made in a gun shop, where markups of fifty percent are routine and markups of greater than one hundred percent are not unheard of. You can usually negotiate the same terms through a pawn shop. If you are very lucky, you might have a federal

firearm licensed friend who will order you a rifle at cost and thus save even more. If you can't find anyone willing to order a rifle for you at cost plus ten percent or so, write to Shotgun News at the following address and ask for their list of FFL holders who will receive orders for you:

Shotgun News
Snell Publishing Company
P.O. Box 669
Hastings, NE 68902

They publish a list yearly of FFL holders who will perform this service and they also state what percent the FFL holders charge for the service.

The two most common "newspapers" for advertising military rifles for sale are Shotgun News (address above) and Gun List (700 E. State Street, Iola, WI,54945). Many shop owners prefer to use Gun List because its offerings are alphabetized under specific headings, making a particular firearm easy to locate. Shotgun News doesn't have much organization making specific firearms harder to locate; that is, you have to spend a lot of time reading through the paper to locate precisely what you want at the best price. However, my experience has been that those selling firearms through Gun List are more like private owners who expect to get retail or better prices for their firearms, whereas advertisers in Shotgun News are more closely associated with wholesalers attempting to obtain a dealer base, and their prices are much more reasonable.

When you are picking out a rifle from the advertisements, look for a statement like "select-add $10" or "Hand pick-add $20". What this means is that the supply of your particular rifle comes in a large variety of conditions. If you order at the base rate, they will grab a rifle at random from the existing supply and ship it to your FFL holder. It may be in very good shape or in not-so-good shape. If you add to the base rate the amount they ask for "select", they will hand pick from existing supplies the best of the lot to send to you. Do it. Believe me, it is worth the extra money.

The most important thing in ordering your rifle is that you get one in which the bore of the rifle is not pitted or the rifling worn out. Most of the time wholesalers have a return policy which says if purchaser is not satisfied with the rifle he can return it within 5 to 10 days (the actual inspection period varies from wholesaler to wholesaler) and obtain a refund. This would cost you return shipping expenses, but sometimes it is better to return a rifle and select another, even if it means eating the return shipping costs, because a bad firearm is generally a lost cause and will never be useful to the purchaser. I have had to return firearms on occasion, but it has been infrequent.

When one receives a rifle, the first thing he should do is examine the bore. If it looks very rough or cruddy, clean it with a bronze bore brush and a good bore solvent. Often military surplus rifles have grease in the bores (and throughout the rest of the rifle as well) to protect against rust during long storage. The grease is usually a thick, dense, brown gummy stuff called cosmoline, but I have seen some Chinese rifles preserved with what looks like regular old black axle grease. As long as a barrel has grease in it, you can't really tell how good or bad it is. If the bore still looks rough after a thorough cleaning, send it back to the wholesaler. Sometimes it helps to ask your FFL holder when ordering to specify that the barrel must be in very good to excellent condition with no pits so the wholesaler will know ahead of time what will be acceptable to the customer and what he is likely to have returned.

Another thing which ideally should be done is to have

the rifle headspaced. Headspace has been defined as "The linear endplay of a cartridge in the chamber with the bolt closed (Speer Reloading Manual Number Nine, Speer, Inc., 1974)." Excessive head space is dangerous because it can cause the whole rear end of the cartridge to blow off causing serious damage to the rifle and likely the shooter as well. A small amount of headspace is tolerable, but if headspace is excessive it does pose a significant risk to the shooter. If your FFL holder is a gunsmith, he probably has a set of headspace gauges for your rifle and can check this for you. Some wholesalers check the headspace on rifles they are selling to ensure they are within acceptable tolerances. Probably most do not. If you have no way of telling whether the headspace is excessive or not, a rule of thumb can be used: if the barrel is in very good to excellent condition and the stock is in pretty good shape, it is probably a good bet that the rifle has not been fired a great deal and the headspace is probably o.k. Obviously, the key word here is "probably", and "probably" doesn't mean "certainly". Another method of checking for excessive headspace is to use a dummy cartridge (one which has no gunpowder in it). Remove the extractor from the bolt, and put the dummy round in the rifle's chamber. The bolt should close easily on the dummy cartridge. Now cut a shim (.006 inch thick) slightly smaller than the diameter of the cartridge rim, and place it on the head of the dummy round. The bolt should not close on the round this time. If it does, you have excessive headspace. Of course, once a round is fired by the rifle in question, you could look for the tell-tale signs of excessive head space in the spent cases, such as excessive stretching of cases and signs of impending case head separation. Target shooters, handloaders and gunsmiths will be able to identify these signs from fired cases if the reader is unfamiliar with them.

Of course, the problem with this method is that it requires firing the rifle with live ammo, a decidedly risky operation if headspace happens to be truly excessive. The best advice here is to say that if one is in doubt as to the headspacing of any given rifle, find a gunsmith and have it checked out.

If you do have the rifle headspaced and the clearances are found to be excessive, you have essentially two choices open to you. The first is to have a gunsmith remove the barrel, have the shoulder of the barrel set back to remove the excessive headspace, and have the barrel reinstalled on the action. Ask your local gunsmith what the charge will be for such a service before having it done. The second way to deal with the problem is to send the rifle back to the wholesaler for refund or replacement with a statement that the rifle is defective.

Once you have received your rifle and have determined that the barrel is in good condition, there is one more job that must be done before beginning the conversion. The rifle must be torn down completely and all preservative grease removed. Cosmoline is an excellent preservative, but it gets very gummy after sitting for a long time and can really be a serious pain to remove. If the rifle was packed in axle grease, this will be easy to clean up. The best thing to clean this gunk off with (both cosmoline and axle grease) is pure mineral spirits, sold at Wal-Mart as paint thinner for less than two dollars per gallon. Gasoline actually cuts the grease better than mineral spirits does, but gasoline is not recommended because of the tendency to barbecue yourself during cleaning if a spark is generated somewhere.

The important thing to remember in the cleaning process is that the entire rifle must be stripped and cleaned. This includes the stock (both action and barrel channels if these contain grease) and it also includes complete disassembly of the bolt, so that the firing pin, firing pin spring, and its housing within the bolt can be cleaned out. This is a fairly simple project, requiring no more than a screwdriver, or sometimes a coin. British SMLEs require a special tool to disassemble the firing pin from the bolt, so you won't be able to take these bolts apart. A tool can be made which will perform this function, however. See appendix B for a description of this tool. Other than the SMLEs, all other military bolts are very straightforward and relatively easy to strip.

If the cosmoline or preservative is not completely removed, a variety of problems can occur. Leaving dried grease in a barrel results in a barrel obstruction. Most barrel obstructions result in either of two things when the rifle is fired. Either the barrel will bell or bulge (ruining the barrel), or the barrel will explode. I must admit that I have never seen a barrel blown due to cosmoline obstruction, but I have seen them blown by a variety of other obstructions and I have no doubt that cosmoline could do it too. When the rifle is fired it should have absolutely nothing in the barrel but the cartridge; not cosmoline or oil, not dirt or mud, not snow, not bugs or beetles, nothing! And above all, the reader should never under any circumstance attempt to remove a barrel obstruction of any sort by shooting it out with a live cartridge. When a barrel blows, shrapnel goes everywhere. It is like a grenade exploding right in your face. Obviously this does not "bode well" for the health of either the shooter or any bystanders that happen to be close.

Leaving cosmoline in the barrel channel of the stock will produce problems when one is ready to bed the barrel and action, a process that will be explained later. Cosmoline left on the interior working parts of the rifle will create problems in a very dirty, gritty environment in that the grease holds grit to the metal parts, and as they are used over and over, the grit acts just like sandpaper in the action, chewing away metal.

Cosmoline inside the bolt may result in a slow rate of firing pin drop producing misfires or worse. Many years ago I bought a military surplus Mexican Mauser; I was a graduate student at Colorado State University at the time, and I didn't have a lot of money, so I picked this particular rifle up as surplus at a Coast-to Coast store in Fort Collins for twenty-three dollars. It appeared to be a sound rifle, so I loaded it with a full magazine of 7 mm ammo intending to determine how accurate it was. The first round went off fine, but none of the succeeding rounds would fire. All had the primer deeply indented, but they would not fire just the same. I couldn't figure out why it wouldn't fire since everything seemed to be working correctly. I finally took it to an ex-Navy armorer by the name of Standring (owner-operator of the Mountain Armory in Fort Collins) and he figured out what the problem was. The inside of the bolt was packed with cosmoline (I had already cleaned the rest of the rifle). When the first round was fired, the recoil pushed back against the firing pin. Normally, this simply pushes the spring loaded firing pin back into its housing in the body of the bolt. But this time the bolt was filled with old gun grease which created resistance to the firing pin being pushed back. The resistance was just enough to cause the firing pin to bend slightly. The bend was enough to shorten the distance of firing pin travel to where it would not fire live cartridges, but not enough to prevent the firing pin from indenting the primers. Problems like this can drive you up the wall trying to figure out why a rifle that appears to be working right doesn't want to fire. So clean all of the preservative from the rifle, inside and out, and you will be ready to start the conversion.

This book will describe the disassembly and bolt stripping procedure for a limited few rifles (see appendix A), but it cannot hope to cover all of the various models on the market at the present time. If the reader has a rifle that he wishes to convert which is not covered in this text, he can find disassembly instructions in one of the following books (which can be obtained through your local library, through interlibrary loan, or can be purchased):

The NRA Guide to Rifles & Shotguns
National Rifle Association, pub.

Firearms Assembly or Disassembly,
Part 4, Centerfire Rifles
by J. B. Wood (1991)
DBI Books, Inc.
Northfield, Ill.

The Encyclopedia of Modern Firearms,
Parts & Assembly, Vol. I
by F. Brownell & Son, Inc.

Any of the above books can be purchased from Brownells, Inc., 200 South Front St., Montezuma, Iowa 50171. Borrowing one of these through the library is a good, inexpensive way to learn how to disassemble your rifle, but buying one or more of these is a better idea, since you will probably need to refer back to it from time to time for the rest of your life.

In my estimation, every person that owns a firearm should be able to tear it down and reassemble it. Especially in a survivalist (i.e. self-reliant) situation, one is not going to have a gunsmith at his beck and call every time a part breaks or a jam of some sort occurs. There are truly some problems that only a gunsmith can repair, but the owner should at least be able to diagnose the problem himself, and can often repair the problem, as long as he understands the mechanics of the firearm.

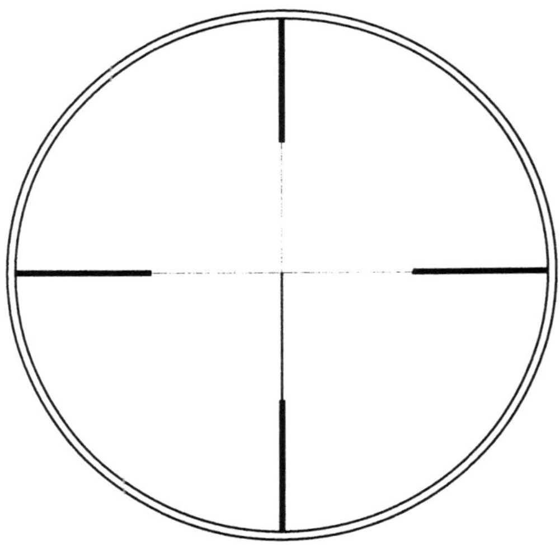

Chapter 5
Converting the Military Surplus Rifle

Basically, there are five areas of concern in rifles which have a definite effect on accuracy. These include the condition and type of barrel, the ammunition one is planning on using, the stock to barrel/action fit, the trigger assembly and the sights system. Each of these areas will be discussed in turn, with a description of how to accurize at each step.

Section I
Condition and Type of Barrel

As mentioned earlier, the type of barrel one gets on any given military rifle varies from model to model. The Finnish model 28 Moisin Nagants have heavy barrels made by Sako which certainly meet match shooting specifications. The best barrels for accuracy are heavy barrels (i.e. barrels with very little or no taper). Whether a rifle has a heavy barrel or not, machining tolerances are rather critical when it comes to accuracy. Unfortunately, how close a manufacturer sticks to machining tolerances when making a firearm also varies from one manufacturer to another. Some are very lax in what tolerances are allowed, and may thus produce bores too large for the caliber bullet the rifle is supposed to shoot. Also, since military surplus rifles are used, there may be enough wear that the bullet no longer fits the bore of the barrel. I have seen some bores so worn that you could drop a loose bullet (that the rifle is bored for) all the way though the barrel and have it fall out of the action. The bullet should be a tight fit and should obviously not slide through the barrel. If it does, the lands are not engraving the bullet. On the other hand, some manufacturers have built their reputation on strictness in maintaining manufacturing tolerances (the Finns are a good example), and are generally reliable when it comes to close tolerances.

Another consideration when selecting a military rifle for sniper conversion is the precise time that the rifle of interest was made. Pre-war manufactured rifles are typically the best. During the height of WWII, as has been the case in every war, the demand for arms was very high. In the manufacturers' attempts to turn out excessively large numbers of firearms to meet wartime demand, requirements for close tolerances and good fit & finish on the military firearm were usually relaxed. As a result, these late war production rifles generally do not look as good as earlier production and typically do not shoot as well either. In the case of German made rifles from WWII this is especially critical, since rifle manufacture for the Third Reich near the end of the war was done by slave labor. As one might expect, slaves do not take well to the yokes of their masters, and some of the rifles made in this way were sabotaged, intended to self-destruct and take its shooter along for the ride. Therefore, if you have a choice between purchasing a pre-war manufactured rifle (or even a post-war manufactured rifle) or one that was wartime manufacture, pass on the wartime manufactured item.

Assuming that you have ordered and received a military surplus rifle, how can you tell if the barrel is bored to the right specifications? There is really only one way to do this.

1. Open the bolt and remove it from the rifle's action.

2. Run an oil soaked swab through the barrel to liberally lubricate the bore.

3. Take a slightly oversized slug of pure lead, and with a dowel rod and a mallet, drive the slug into the barrel. If you can't come up with a pure lead slug, you can use a pellet of 00 buckshot.

4. Using a long dowel rod, tap the slug all the way through the barrel until it drops out the action.

5. Retrieve the slug. The barrel has spirals in it which gives fired bullets their spin, stabilizing them in flight. The raised areas in the barrel are known as "lands" and the low cut areas of the barrel are known as "grooves". You will see that the lands of the barrel have cut grooves in the slug. These grooves in the slug now correspond exactly to the rifle's lands and the measurement across the raised portion outside of the slug's grooves correspond exactly to the barrel's groove measurements.

6. Using a caliper accurate to at least one thousandth of an inch, measure the diameter of the slug both outside the grooves to get the barrel's groove measurements and within the grooves to obtain the barrel's land measurements. If you don't have a caliper capable of measuring with such fine accuracy, you can take the slug to either a gunsmith or a machinist, and they can make the measurements for you.

Once you have these measurements, you have something with which to assess wear and usefulness of the barrel. You do this by first looking at the groove diameter and comparing it with the diameter of the bullet it is intended to shoot. The bullet intended for use in American military rifles is .308 inch in diameter. Ideally, the groove measurement should be the same, i.e. .308 inch in diameter. If the groove is much greater than .001 inch oversize, it will normally not function adequately as a sniper rifle. Of course, there are exceptions. The Japanese Arisaka is usually loaded with a cartridge containing .264 inch diameter bullets, but the rifles' barrel grooves are generally bored to .266 or even larger (de Haas, 1984), yet they usually perform marvelously. If one is a reloader or knows someone who is a reloader, he can compensate to some degree for slightly oversize barrel grooves in .30 caliber rifles by using a .309 inch diameter bullet. A few manufacturers make a bullet in this .001 inch oversize version for match shooters. The oversize portion will take up the extra space between the bullet and the oversized barrel, but this will only work for slightly oversized barrels.

The problem with oversized barrels is that hot gases from the fired cartridges blow out around the bullet as it travels down the barrel. This always results in a drop in muzzle velocity, and typically further erodes the grooves. The larger the gap between bullet and barrel, the more barrel erosion one may expect and the lower the bullet velocity.

The next thing that needs to be examined is the height of the lands. Lands on newly manufactured barrels generally range from about .002 inch to .006 inch. If the lands are very shallow (i.e. less than about .0005 inch) then you have two problems. The first is that you have very little barrel life left. Wear is to the point that before much additional use, there will be no more lands. The second problem is, when the lands wear down to where they can't get a good "bite" on the bullet, the bullet "skids" down the barrel instead of getting a good twist on it. When this happens, the stabilizing spin on exiting bullets varies greatly, and you get shots which are aimed to the same point hitting all over the place. Accuracy is gone in this case, and no amount of accurizing will help it. Once again a rule of thumb can be used in evaluating whether existing lands are of sufficient size to warrant use. Hold the barreled action (with the bolt removed) up against a lighted background. If the spiral grooves look very strong and prominent, the height of the lands is probably sufficient. If no spirals are seen, or if they appear very faint, the wear on the lands is excessive. The bad part of this rule is that it takes a certain amount of experience to use. By this I mean that a person must have peered down the bore of quite a few rifles in various conditions in order to accurately arrive at a conclusion of what the terms "strong and prominent" mean. Obviously, if there are no lands left, or only faint traces of lands, anyone can discern this. It is the "in-between" conditions of the barrel that are difficult for the novice to evaluate. If the means to make micrometer measurements are available, it is probably better to do this, as it will leave nothing to question.

If your newly received rifle has inadequate tolerances for either lands or grooves, I recommend returning the rifle for another. The only real solution to this kind of problem is to have the barrel removed and replace it with another one. Asking a gunsmith to perform this service for you would cost an additional

$100+ (assuming a new barrel is ordered rather than a used one).

Pits in the barrel should be avoided at all costs if you intend to get sniper accuracy out of a rifle. Pits are nothing more than many tiny holes in the bore of the barrel caused by corrosion. Although environmental conditions can cause pitting from rust corrosion (such as leaving the rifle in an outdoor shed for a long period of time where it is exposed to excessive dampness), it is usually due to the use of corrosive primers. Many years ago, most military cartridges utilized primers which contained perchlorate. When detonated, the perchlorate priming compound released chlorine, an extremely corrosive element. If, immediately after shooting, the bores were cleaned thoroughly with soapy water, followed by dry patches, followed by a bore solvent cleaning, followed by oiling the bore, corrosive pitting would not occur. However, in wartime conditions troops were often not as meticulous in cleaning their rifles as they would have been otherwise, their time being taken up with more mundane matters (such as keeping their head down and returning fire) and corrosive pits in the barrel were often the result. It is doubtful that a heavily pitted barrel will ever have the innate accuracy needed for a sniper rifle. However, it seems there are always exceptions. In 1976 I bought a model 94 Winchester rifle (with a 26 inch barrel, full buckhorn open sights, and chambered for the .32 Winchester Special cartridge) from an elderly gentleman. The first thing I always do when I buy a used rifle is to clean the bore. In this case, I scrubbed for three days before the bore was clean enough to suit me, and even then it was obviously badly pitted. Shortly afterward, I asked the gentleman who sold the rifle to me when he originally purchased the rifle. His reply was "1928". I then asked him when he had last cleaned the rifle; his reply to this was "Oh, I've never cleaned the rifle since I've had it!" So I had the corrosion and pits of nearly 50 years of neglect to deal with. I knew it would never be accurate, but I planned a trip to the range anyhow and gave the Winchester to my Dad, intending to rag him for his poor performance with a "classic" Winchester. You can imagine my surprise when Dad shot cloverleafs at 100 yards, so close together that you could touch all of the holes with your thumb. The moral of the story is that occasionally a firearm will shoot very accurately even with a badly pitted bore, but I must confess that these occasions are the exception to the rule. I have seen more pitted barrels through which you couldn't hit the broad side of a barn than I have examples like that old Winchester.

If the bore of the rifle is not pitted, you can proceed with the sniper conversion. If it is pitted, frankly, I don't think it is worth the conversion effort.

At this point the reader may be gripped in the depths of despair, thinking that with so many restrictions he is never going to find a surplus rifle worthy of conversion. He should not be too worried about this. There are huge numbers of military surplus rifles on the open market, and a fair number have good to excellent bores. If the reader is careful in his selection, he will come up with a good specimen well worthy of conversion.

If the bore of your rifle is not pitted, check the muzzle face for damage. The outside circumference of the barrel at the muzzle can be dented and dinged, but if the dents and dings extend down to the barrel opening, then accuracy is likely to suffer. The reason for this is because the process of denting displaces metal, and when metal is displaced at the point where the fired bullet exits, it frequently throws the bullet off course by a fraction of a degree. Unfortunately, a fraction of a degree off course at the muzzle can (and usually does) mean a very big deviation at 100 yards. This is a problem that is easily repaired by lopping off a short piece of barrel, dressing it up with a file and recrowning it. This is the same procedure used to shorten a barrel. The process is outlined as follows:

1. Remove the barreled action from the stock.

2. Clamp the barrel in a vise, with the end of the barrel protruding horizontally about three or four inches beyond the jaws. Be sure to pad the jaws of the vise with either heavy leather or wooden blocks to keep from marring the outer surface of the barrel. Also, support the action with wood blocks to take the weight of the action off the barrel.

3. Using a scriber or some pointed tool, scribe a line equidistant around the barrel at the place you want to cut it. If you have a damaged muzzle face that you want to correct, you only need to remove about 1/4 inch of barrel. If you are deliberately shortening a barrel, don't worry about removing the piece which has the front sight attached to it, since a sniper conversion uses a telescopic sight and the front sight is not needed.

4. Using a hacksaw, taking care to keep the cut square with the end of the barrel, cut the barrel off at the scribed line. Use a fine hacksaw blade (32 teeth to the inch). This will make for slow cutting, but the cut will be much finer and will require less work with a file later.

5. Remove the barreled action from the vise, reposition it vertically with the tip of the barrel protruding about two inches above the jaws of the vise. Continue to protect the barrel with leather or wood padding.

6. Using a flat file, dress up the end of the barrel by filing off all hacksaw marks. Make sure that you file square with the end of the barrel. Use alternating 45 degree strokes across the end of the barrel while filing: first a 45 degree stroke to the left across the muzzle, then a 45 degree stroke to the right, then back left, then back right, etc. This will allow your file to remove any marks which might be made by the file itself. The end of the barrel should be cut and filed perfectly square when you are done.

7. Using a file, remove the sharp edge on the outer circumference of the barrel muzzle. How much you may want to round this off is a matter of preference, but it only needs to be rounded off enough to dull the edge so that it won't cut your hands or clothes.

8. The last step involves crowning. The barreled action doesn't have to be in the vise to do this. You will have to have a tool made for you. Find a ball bearing about twice the diameter of the bore (a half inch diameter bearing will work fine), take it to a welder and have him braze the ball bearing onto the end of a piece of straight metal rod about 6 inches long. This is the only tool you will need to crown the barrel. Get a small container of valve grinding compound from the local automotive parts house. Clover brand is a good choice; this has a double sided container which has a coarse grade of grinding compound in one side of the container and fine grade in the other side. The grinding material contains both grit and grease mixed together.

9. Chuck the tool into an electric hand drill. Dip the ball bearing into the coarse grinding compound, hold the ball bearing to the barrel opening and grind by using the slowest drill setting possible. Squeeze the drill trigger until the tool begins to spin, then release it,

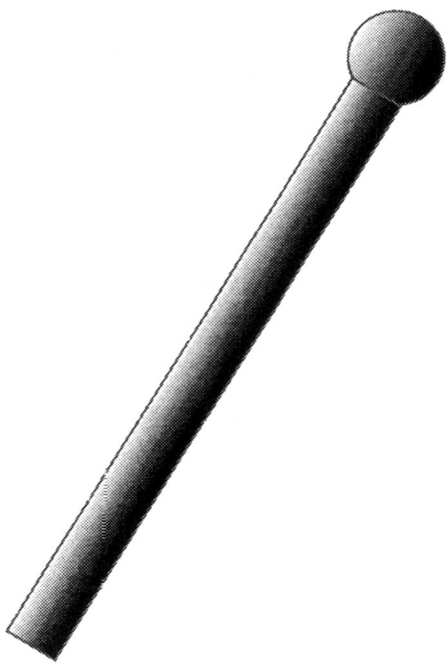

A half inch diameter bearing welded to a piece of straight metal rod about 6 inches long is the only tool you will need to crown the barrel.

Section II
Ammunition - Selecting for Accuracy

This section is primarily intended to advise the reader on ammunition choices for obtaining the best accuracy with his modified rifle. Although most people probably don't think a lot about ammunition, and most probably assume that any given rifle should be about as equally accurate with any ammunition used, they would be wrong. Just ask any Marine Corps sniper, match shooter or varmint hunter. The rifle is one third of the accuracy problem and the ammo used is another third (the shooter himself makes up the last third). In order to obtain very fine accuracy, you need to optimize both rifle and ammo.

The current Marine Corps sniping manual specifies two acceptable types of ammunition to be used with the 7.62 NATO (.308 Win.) chambered sniper rifle. The preferable ammunition is M118 match ammo, shooting the 173 grain boattail bullet. The head of the bullet is marked "L.C." for Lake City arsenal where the ammo was produced and "match" indicating match quality ammunition. If this is not available, M80 or M80E1 military ball ammo using a 147 grain bullet may be substituted, but with the warning that it will not be as accurate (hits higher on the target). The manual states "Match quality ammunition will normally be issued because of its greater accuracy and reduced sensitivity to the wind. However, if match ammunition is not available, a different grade of ammunition may have to be used...Lake City M118 match grade ammunition will be issued to the sniper and should be fired at all times, when available (p. 2-9)."

It is difficult to see where we could improve on this. Match grade ammunition is without a doubt the best quality made and for sniper applications it should be used over any other type of round. Fine accuracy for the sniper round comes from several components. These include :

1. Consistency in powder charge from round to round.
2. Consistency in primers.
3. Type and weight of bullet.

Match ammunition is more meticulous when it comes to accuracy than any other type of manufactured ammo. It has to be, since it is used by international competitors for target shooting. Quality control for match ammunition ensures that powder charges are thrown with very little deviation from round-to-round. Hunting loads do not get the same careful treatment, probably because in a hunting situation, the hunter is usually going to be shooting for the heart/lungs of a deer or elk, and being off by an inch is not going to make a difference. In international competition, however, it may mean the difference between winning and losing.

Consistency in the use of primers is very close in match ammunition also. Different brands of primers will produce different velocities by themselves and a difference of a few hundred feet per second can affect bullet grouping. I believe manufacturers of match ammo only use a single primer for their products to ensure consistency. Hunting ammunition is not so carefully controlled. I have, on occasion, bought factory hunting ammo with part of the cartridges' primers brass colored and part of them silver colored, obviously not the same lot of primer and probably not even the same manufacturer. It has been many years since I have seen this, but I have never seen it with any match ammo.

Type of bullet has a large effect on accuracy also. There are many types of bullets on the market for a variety of applications. There are flat nose bullets for rifles with tubular magazines, round nose, spitzer (pointed), soft points (the tip of the bullet has lead exposed), full metal jacket (no lead is exposed), hollow points, brass jacketed, steel jacketed, flat base, and boattail. This lists only a few of the possibilities out there, and all of them usually have some effect on accuracy. To make a long story short, both flat and round nosed bullets are not used in sniper rounds. They are ballistically inefficient. They have other applications, but sniper is not one of them. The spitzer point is preferred; whether it is soft point or not doesn't matter, although military ammo is uniformly full metal jacketed. Steel jackets are best avoided if possible because it wears the barrel of the rifle quicker than brass jackets do (you don't usually find this in any but imported foreign ammo or ammo used for hunting African big game). Best accuracy is obtained from boattail bullets. These are bullets that have a bevel cast into the base of the bullet. You wouldn't think that the base of the bullet would have any effect on accuracy, but it does. Boattails increase the ballistic efficiency (think of this as accuracy), sometimes considerably.

Bullet weight also has an effect. Lighter weights (e.g. 140 to 150 grains) are generally more accurate, but they are much more prone to being blown off course by the wind. Very heavy bullets (in the 200 grain or over range) are affected the least by wind, but are not nearly as accurate. The optimum is around 165 to 175 grains of weight. Hornaday manufactures an excellent 168 grain hollow point boattail match bullet for .308 bores which I can recommend heartily as a sniper bullet.

If match grade ammo is available, use it. If it is not available, the next best alternative would be handloads. In fact, benchloaders would probably tell you that handloads are preferable to even match grade ammunition. As a general rule match shooters are also handloaders, because by controlling the manufacture of their own ammunition, they can customize it to their own satisfaction. Rifles are like people in that every one has a personality of its own, each with its own quirks and peculiarities. It is possible to have two rifles of the same make and model, in approximately the same condition, and have them shoot totally different patterns with the same ammunition. Target shooters realize this and spend quite a lot of time experimenting with various powders, charges, primers, bullet types, etc., until they find the combination which provides the ultimate in accuracy for their own particular rifle. The catch to this is that it takes quite a lot of time and effort (but it is a lot of fun!). If you are not a handloader, but you have a friend who is, you may be able to get him to help you work up a load that is extremely accurate for your rifle. Once you have the accuracy "formula", you can have any handloader make up a batch for you anytime you need it, or with the proper training and a few basic hand tools, you could do it yourself.

The third best choice for ammunition would be military surplus ammunition of fairly recent manufacture. Ammunition that is extremely old (i.e. older than about 30 years old) may produce misfires. In fact, depending upon how old the ammo is, it may produce a lot of misfires. There is no way to predict how old is too old, but I would be suspicious of anything greater than 30 years old. I am probably erring on the conservative side here, but when you are in a firefight is not the time to find out that every other round in your batch of ammo is going to misfire. Old ammunition is very hard to second guess. I got a case of 1960s Egyptian manufacture 9 mm parabellum a few years ago. When the ammunition lot arrived, the cartridges all fired fine. Within about 8 months, the primers had gone bad in the entire lot. I kept the ammo in a cool dry storage area, so I don't have a clue as to why it went bad. I ended up pulling the bullets so I could salvage at least that part of the shipment, but cases were berdan primed (a European priming system not compatible with the American boxer primers) so I couldn't even salvage the cases.

Military surplus ammunition is generally very cheap compared to commercially made ammo. This makes it desirable for target shooting. It is generally not usable for hunting because military bullets are all full metal jacketed (fmj), whereas hunting rounds are required to have a lead core and have lead exposed in the nose of the bullet (so the bullet will expand in game and be more likely to provide a clean kill). Also, some military ammunition has a steel core rather than lead. Military ammunition which is older than 1980s manufacture usually has another drawback also. It is corrosive and can pit the bore if the rifle is not cleaned promptly after firing. It is better to avoid corrosive ammunition when at all possible, but it can be used if the bore is cleaned IMMEDIATELY after use with hot soapy water, followed by dry patches to remove the water, followed by cleaning with a good bore cleaner (such as Hoppe's Number 9, or milspec cleaner which contains no grit), followed by swabbing with gun oil.

In spite of all of the drawbacks, military surplus ammo is preferred over hunting ammunition. Military ammo is supposedly loaded to certain fixed specifications using large lots of the same powder. This results in greater consistency than is generally found in hunting ammunition.

If you have no alternative, hunting ammunition is always a last resort. The reason it is a last resort is because you never know what powder or charge is going to be used in commercial ammo. Manufacturers use whatever powder they have on hand at the time and supposedly load to some fixed velocity, but the truth is there is a lot of variability here. Bullet weights will be fairly consistent as will cases, but there may be considerable variation in about everything else. Make no mistake here, I'm not saying that regular hunting ammunition is inadequate for normal hunting at normal ranges. It is fine for this application. I ought to know, I've taken enough game with commercial ammunition. I'm just saying that it is not the best choice for sniper applications where the shooter needs fine accuracy over very long distances.

The order of choice changes somewhat if the surplus rifle you are converting is chambered for the 8 mm Mauser. If you are not a handloader, you should use either Norma ammunition (Swedish made hunting ammunition marketed in the U.S.A.) or military surplus ammunition. The reason for this is because American manufacturers grossly underload the 8 mm Mauser. It is generally loaded to muzzle velocities of about 2350 feet per second (fps). Perusing just about any handloading manual will show you that the cartridge could be safely loaded (with lighter bullet weights) to at least 2800 fps, and in some cases closer to 2900 fps. In fact, in the 1994 Gun Digest, Jim Thompson states that American manufacturers underload the 8 mm Mauser so much that European publications list it as a whole different caliber (p. 18). Norma ammunition is European ammo, and is loaded up to its capabilities. Of course, military ammo is also loaded up to its full potential.

The main thing to remember is to be consistent. If you use commercial ammo, use the same brand in the same bullet weight. If you handload, use the same powder, the same powder weight, same primers, same cases, same bullet type and weight. CONSISTENCY always produces the best results.

If you are converting a rifle and you are interested in determining how much improvement the accurizing process has generated, make sure that you fire it before you do the conversion. This will give you a standard for comparison. Make sure that you use the same ammunition for both pre and post testing. If you change bullet weights, powders, brands, etc. between pre and post testing, you won't be able to tell if differences in accuracy are due to the rifle's accurizing modifications or due to differences in ammo. If you are only interested in getting the most out of your rifle, then post testing is all that matters and the pre testing is superfluous. In this case, just do the accurizing steps and do your test shooting later.

Section III
Stock to Barrel/Action Fit

At this point we are ready to begin the actual modification of the rifle. At the time of writing this manuscript, I was given a KAR98 rifle, a military surplus 98 Mauser, that seemed on the surface to be ideal for the conversion. However, a cursory examination showed the bore to be severely pitted, not really suitable for a conversion to sniper configuration. A decision was made to use it to demonstrate barrel free floating (described later) as this could be done without making any permanent changes in the stock. The condition of the barrel was such that a barrel replacement was called for, a subject too technical for this manual.

In order to prove the point of this manual it was decided that I would start from "scratch", that I would purchase a military surplus firearm and make a direct conversion in the same way that I am advising readers to do. I settled on using a number 4 mark 2 Short Magazine Lee-Enfield (SMLE). I found a supply of these rifles (obtained through Interarms Corp.) at Nationwide Sports, 1460 Linda Way, Sparks, Nevada. Nationwide wholesaled the SMLEs out to dealers for $61.90 for one rifle, or $59.90 each in lots of 3. What made their advertisement interesting was the description of the rifles as being in "NRA very good to excellent condition" and the added note "Sorry, no hand picks!" which I took to mean that all were in about the same general condition, i.e. exceptionally good. Incidentally, at the time of this writing it is possible for one to purchase SMLEs in the number 1 mark 3 model for as low as $39; these are Australian manufacture and are in a poorer condition than either NRA very good or NRA excellent, but it goes to show how inexpensive military surplus rifles can be. Moisin Nagant rifles are available for as low as $34.90 in NRA very good condition. Remember that the prices quoted here are wholesale, and that it

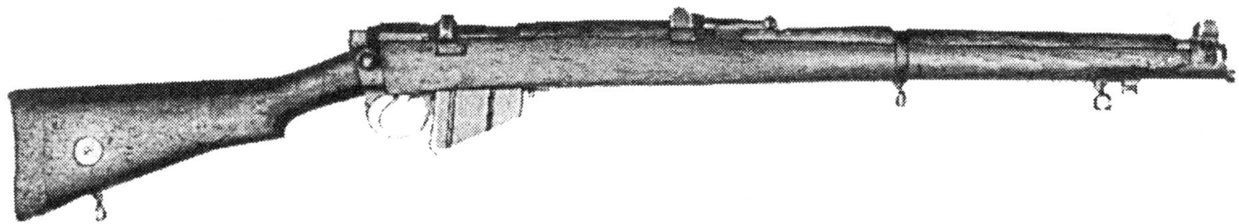

An example of the extreme low cost of some military rifles is SMLEs in the number 1 model for as low as $39; these are Australian manufacture and are in a poorer condition than either NRA very good or NRA excellent.

will cost the reader about 10% more if he negotiates his prices as described earlier.

At any rate, I ordered the number 4 SMLE in, and when it arrived I found the "NRA very good to excellent..." statement to be rather loosely interpreted by the wholesaler. The stocks did not appear to have been used extensively, but there were a lot of dings in the wood, probably the result of much handling over the years since the rifle was manufactured. The rest of the rifle was packed in enough cosmoline to grease an army tank. Gun grease was everywhere.

The first task was to completely dismantle the rifle and remove the grease . Once the grease was removed, the external metal of the rifle proved to be in reasonably good condition. By this I mean that the metal was not worn, but it was not pretty either (it is important to realize that even an "as issued" military rifle manufactured during conditions of war as the number 4 SMLE was, will not be very pretty. The finishes are usually rushed and can look rather rough compared to

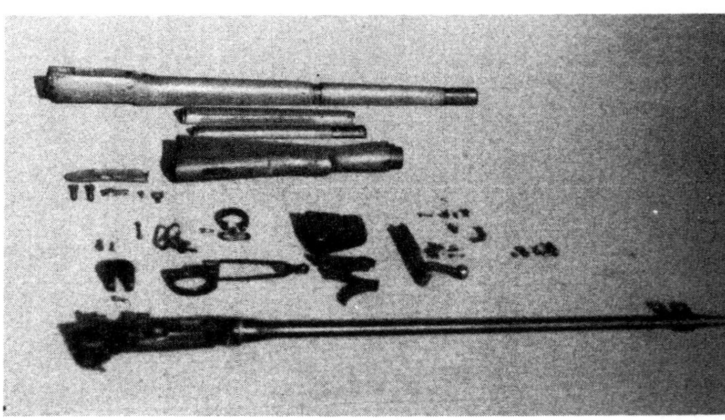

A completely stripped No. 4 Enfield Rifle. Complete stripping is necessary to clean the cosmoline on many surplus military rifles.

The Short Magazine Lee-Enfield (SMLE) Rifle No. 4

commercial finishes). Of course the important thing was not external appearances, but the condition of the barrel. When the cosmoline was cleaned from the barrel to where it could be inspected, it was found to have a bore that looked to be unfired. There was no evidence of any pitting. I drove a 00 buckshot through the barrel and measured the land and groove diameters. The land diameter measured .303 inch and groove miked .3115 inch, which lead me to believe that the rifle was indeed unfired, or at least had been fired very little. These are exceptionally good measurements for an SMLE. Land measurements on average SMLEs are generally around .303 inch and groove measurements are usually around .314 inch. When the Lee-Metford (SMLE's forerunner) rifle was first put into service, the British immediately got complaints from the field that rifle bores were eroding very rapidly. This was probably due to a couple of factors. First, the British ammunition used cordite, a smokeless powder extruded in long strands like spaghetti, which was cut to the full length of the cartridge cases. This type of powder proved to be very hot. The other factor was the excessive clearance between the bullet and the barrel's groove diameter. Since the bullet was .311 inch diameter, this meant there would be as much as .003 inch clearance, and such clearance is never good. The obvious solution was to bore barrels with groove diameters of .311 inch and go to a different type of powder for their ammunition. For whatever reason, the British decided to take a different tack. They compensated for the erosion by producing barrels with very broad and unusually deep lands, thus you end up with lands about .011 inches high. Although their solution to the problem was not one I would have chosen, it seemed to work. The barrel life increased without a concomitant loss of accuracy. If you are lucky enough to get an SMLE with a groove diameter near .311 inch, consider yourself fortunate; you are just that much ahead of the game. The bullet whose diameter is equal to the groove diameter of the barrel will produce higher velocities and generally result in a slower rate of barrel erosion. Although Hornaday makes their .303 British bullets in a .312 inch diameter, other manufacturers don't. It is unfortunate that bullet manufacturers don't make the .303 in standard oversize diameters of .312, .313 and .314 inch, allowing handloaders to take advantage of bullets which precisely fit the groove diameter of their particular rifle.

At this point, one other idiosyncrasy needs to be discussed where the SMLE is concerned, and that has to do with chamber diameters (i.e. the chamber within the barrel which holds the cartridge while it is being fired). DeHaas (1984) points out that although the earlier number 1 SMLEs have normal maximum chambers, the number 4s have chambers well in excess of this. The reason was to allow the rifle to function under extremely dirty conditions. The excessive chamber diameters do not make the rifle unsuitable for either military or sporting use, but it does result in pronounced cartridge case body enlargement which can cause problems for handloaders. If the metal in a cartridge has to be worked very much (as is done in the full length resizing of cases which have been blown out to excessive dimensions), it will result in early case failures, sometimes after only two or three uses. Perhaps this is the reason that one rarely finds used .303 British brass on the market as you do with so many other cartridges. What to do? If you want closer chamber tolerances, your only alternative is to rebarrel. If you are willing to live with standard military tolerances, deHaas has suggested several ways around the problem:

1. You could run the fired cartridge case into the resizing die just enough to permit you to chamber it in the rifle with a slight amount of effort. Unless you have a lot of patience, this is probably not too desirable.

2. You could get one of the manufacturers of reloading equipment (such as RCBS, Lee or Lyman) to make you a resizing die tailored to the rifle's chamber. You do this by sending about a half dozen fired cartridge cases to the company; they will measure the fired case dimensions and make a die which duplicates the chamber of your rifle. If you decide to do this, call the company first and clear it with them. I know that RCBS will do this for the reloader, but some of the others may or may not. However, this is a good solution.

3. The last solution would be to get dies with neck size only, leaving the body of the fired cartridge the exact size of the chamber as the case is extracted. A cartridge reloaded in this way would probably not chamber very easily, but would probably be the most accurate of all. Match shooters would undoubtedly opt for this choice. This would be a reasonably good solution.

With all of this discussion about the SMLE one might get the impression that they are not a very good rifle. This would be a mistaken impression, however. The SMLE has gained a reputation as one of the truly fine battle rifles of the world. In fact, Garry James (1987, p. 53) has described the number 4 mark 1 (T) as "...perhaps the best sniper rifle of WW II." and deHaas (1984) states that for the number 1 and number 4

SMLEs "The action is strong, reliable and easy to operate, smoother, too, than many other military turnbolt actions." and "For most big game hunting, few cartridges are better than the .303 British (p.69)."

Problems with Wood Stocks

All military surplus bolt action rifles come equipped with a wooden stock. In fact, all Korean War vintage military rifles and before are fitted with wooden stocks. The first fiberglass stocks were introduced during the Vietnam conflict, and since that time we have seen the introduction of a variety of other exotics, such as carbon fiber and various plastics. Our concern for now is that of the wooden stock.

Wood undoubtedly provides the most beautiful mounts for any rifle, military or commercial. But the wooden stock on the rifle, military or otherwise, produces a few problems which must be overcome to achieve high degrees of accuracy. The first problem has to do with production cutting of the stocks. This is traditionally done assembly-line fashion, meaning there is often quite a lot of clearance under and around the barreled action of the military rifle, especially in the area of the recoil lug. This permitted the barreled action to shift slightly in the stock when the rifle was fired. This slight shifting affected accuracy in varying degrees, sometimes quite a lot.

The second problem had to do with the warping characteristics of wood. If the wood used for stock manufacture was not completely cured before being shaped, or if it absorbed moisture through the inletting, the stock would often warp after the barreled action was fitted to it. In his book *Do-It-Yourself Gunsmithing*, Jim Carmichael (1977) described a problem in his .224

Clark commercially made varmint rifle which is sometimes found in military rifles. He said his rifle was initially fitted by hand and was highly accurate, but over a relatively short period of time the groups began to open up. At the point where he began to get stringing of shots, he knew he had a bedding problem. When he took the rifle apart and inspected the stock, he realized that it had warped up and to the right. There was so much warpage that it took a lot of muscle to force the barrelled action back into the inletting when reassembling. Once forced into place, the rigid barrel held the stock straight, but uneven pressure was applied to the barrel by the stock resulting in poor accuracy. This is one of the problems with wood. Warping can and does exert uneven pressure on the barrel resulting in poor accuracy.

A slightly different situation can produce the same problem. The inletting can be uneven causing the barrel channel to exert uneven pressure on the barrel at some point (especially close to the muzzle). As the rifle is fired, the barrel will heat up from friction and expand. As the barrel expands, increased uneven pressure is generated between the barrel and the bearing point. This usually results in vertical stringing, where there may be very little horizontal variation in shots but a lot of vertical variation, making bullet holes appear to line up in a vertical string.

How do we correct this problem? There are three ways this can be done:

1. The first way is to replace the wooden stock with a synthetic, such as fiberglass or carbon fiber. These stocks have two advantages. First they are very light, much lighter, in fact, than wooden stocks. This is generally seen as an advantage, but may be a

disadvantage if the rifle is a hard kicker. Weight in a rifle serves to soak up recoil, and making a hard kicking rifle lighter will only make it kick worse. The second advantage is that the synthetic stock is utterly waterproof; it will never warp, and is completely even in its channel configuration, partly because of the materials it is made out of, and partly due to the manufacturing process. In other words, it eliminates most of the objections against wood. On the down side, they are utterly ugly in appearance, most being dark grey to black with a crinkle finish, and the rest coming in with a camouflage pattern or a leaf print on the stock. A few are made with wood grain molded in, but I have never seen one that looked like real wood. Still, in the minds of some, the advantages of synthetic stocks outweigh the disadvantages. In fact, one of the main characteristics of Jeff Cooper's scout rifle concept is that of a synthetic stock.

Prices for synthetics vary a lot, depending upon the rifle one wants stocked, the stock manufacturer and the design desired on the stock. Ram-Line makes a line of synthetic stocks which run from about $70 to about $90 wholesale, with stocks for some of the more common rifles (such as the SKS, the Ruger Mini-14, and the M-1 carbine) coming in at around $35 to $38. Bell and Carlson have a fairly wide selection of synthetic stocks, but their prices run from about $100 to about $130 wholesale. The only military rifles that synthetics are made for at the present time are the SKS, the M-1 carbine and the model 98 Mauser.

2. The second way of dealing with the wooden stock problem is to free float the barrel. This is a simple process of removing a very small amount of wood around the barrel to leave no wood touching it at any

point. When this is done correctly, the barrel floats free in space above the forearm of the stock. There should be enough space around the barrel to allow one to slide a single sheet of notebook paper between the stock and the barrel with ease.

3. The third way of resolving the wood stock problem is to glass bed the stock and barrel. This involves preparing the wooden stock channel, pouring a synthetic epoxy resin into the channel, then attaching the barreled action to the stock. When the epoxy sets up, it produces a molded "bed" for the barrelled action for a precise fit, and I do mean precise. It is so precise that the bed will even duplicate marks, scratches and numbers stamped in the metal. When the bedding compound cures out, it is so strong that even if the wood tried to warp, it would be unable to do so. An action which has been glass-bedded cannot move in the stock. Even minute movements are pretty much eliminated, resulting in a very high accuracy potential.

Which of these methods is the best? It depends upon who you talk to. Some swear by glass bedding, others by free floating, yet others by synthetic stocks. Certainly the cheapest and easiest method is free floating and the most expensive method is the use of synthetic stocks. In truth, if one were going to use free floating, I would recommend using a combination technique in which the action and the back two inches of the barrel are glass-bedded, then the remaining part of the barrel is free floated. From my own point of view, I prefer glass bedding both the barrel and action. The reason for this is because once a barrel and action have been glass bedded, they are fixed and immobile. If the barrel is free floated, however, it is still free to move if something falls on top of the barrel. Also, foreign material (crud) is free to get in around the opened barrel channel which could lead to the same problem of uneven pressure on the barrel that free floating was used to eliminate, and the stock is still free to warp until it eventually puts pressure on the barrel again. Glass bedding puts an end to all of these objections.

Free Floating the Rifle Barrel

There are basically two ways in which the barrel of any rifle may be free floated. The first way is the one which has traditionally been used by gunsmiths:

1. Remove the barrelled action from the stock so that you have open access to the barrel channel.

2. The first method for free floating involves the use of scrapers to remove wood from the inletting channel until there is none touching the barrel at any point. Inletting scrapers are available from Brownells, but they are also easy to make yourself from scrap metal. Write Brownells (mailing address given in the text earlier) for their catalog and look at the scrapers they offer for sale. You will see that they are little more than rounded pieces of metal with a sharp edge ground into the tool. The use of scrapers allows the home workman to remove only as much wood as necessary to produce wood/barrel clearance. Using sandpaper wrapped around a dowel rod of approximately the same diameter as the barrel will also work, but not as well. This method will frequently give you more clearance than you want, and generally doesn't produce the craftsman-like finish that you often find when you have the job done by a gunsmith, but it will work.

3. When you think that you have enough wood scraped or sanded to provide the necessary clearance, use a candle to blacken the bottom of the barrel with soot.

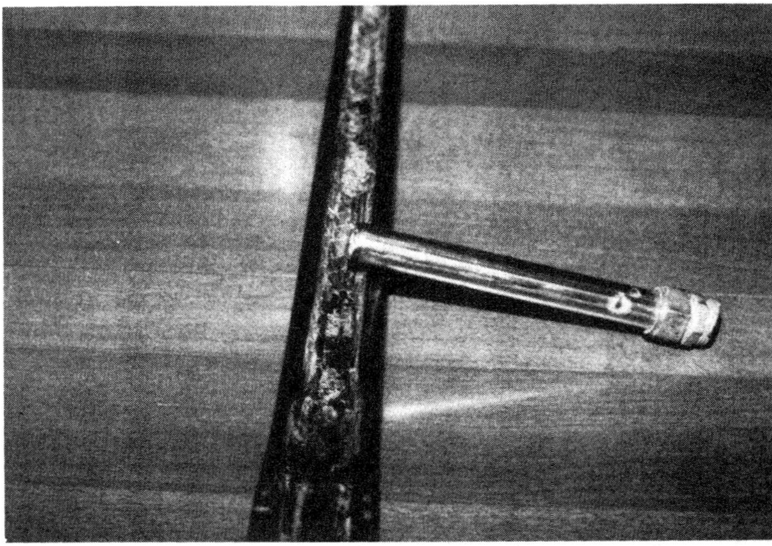

Inletting scrapers are available commercially, but they are also easy to make yourself from scrap metal.

4. Carefully bolt the barrelled action back into the stock, keeping the bottom of the barrel from touching the barrel channel as much as possible until the stock bolts are drawn up tight. Wait for a few seconds, then remove the barrelled action from the stock again. Any soot marks on the wood are indications of areas in which the wood is touching the barrel.

5. Scrape or sand these marks away, re-soot the bottom of the barrel and repeat the procedure. Continue until no marks are left in the barrel channel.

The idea of free floating is to provide only enough space between the barrel and the barrel channel so that a single piece of notebook paper can slide easily between the stock and the barrel. If you can do this, you have met your goal.

The second way of free floating is much simpler,

requires practically no work and uses only scrap material to accomplish the same goal, and is about as effective as the first method.

1. Remove the barrelled action from the stock.

2. Go to a local tire shop that handles truck tires and ask for a discarded truck inner tube that has been trashed. Truck inner tubes fail for a variety of reasons, and discards are common. Try to get one that is smooth, without the ridges that some are made with.

3. Using the action of the rifle itself as a template, cut two small squares of rubber which will fit into the bottom of the recoil lug recess, one square on either side of the recoil lug bolt hole.

4. Bolt the barrelled action back to the stock with the stock bolts. See if a piece of notebook paper will slide

Your local tire shop is an excellent source for free material to use in free floating your sniper rifle.

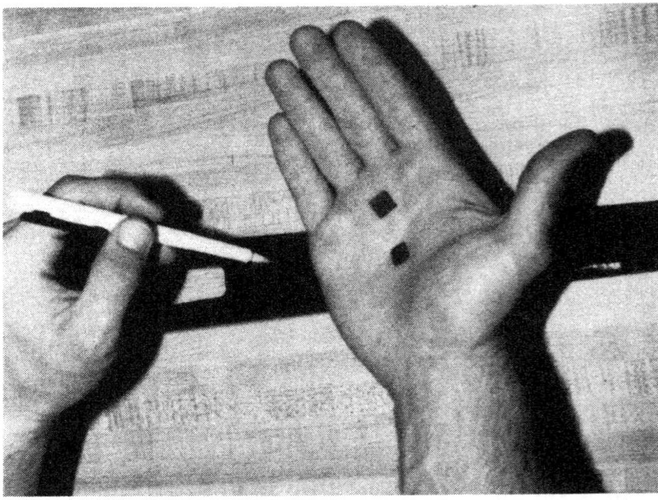

Using the action of the rifle itself as a template, cut two small squares of rubber which will fit into the bottom of the recoil lug recess.

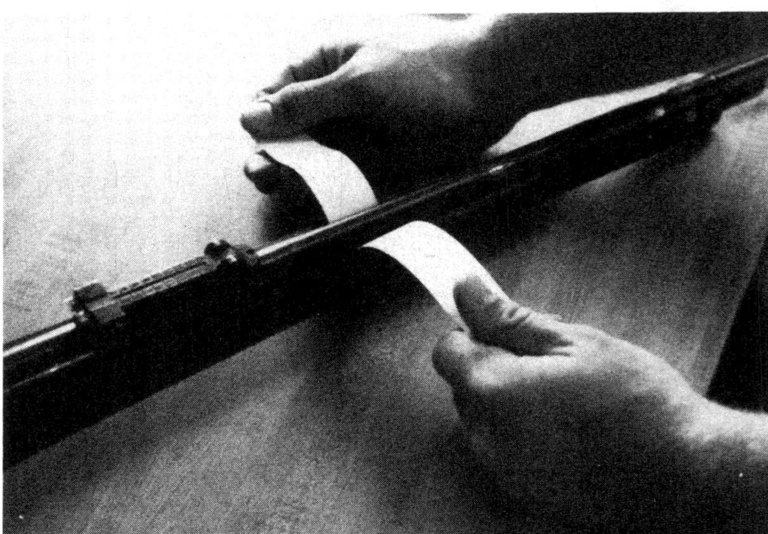

Check to see if a piece of notebook paper will slide between the stock and the barrel. If it will, your job is done and you have floated the barrel.

between the stock and the barrel. If it will, your jcb is done. If it won't, cut another gasket and place it right on top of the first one.

This method is incredibly simple and it will produce results about as good as the first method. Take your choice. I no longer scrape barrel channels unless I have a specific order for it. Some purists will probably want to go with the scraping, but once most people see how simple and cheap the second method is, I imagine they will go this way. The truly best way to use free floating, however is to glass bed the action and about 1 to 2 inches of the barrel forward of where it screws into the action. If you do this, you will have to scrape wood away in the barrel channel, as glass bedding provides zero clearance and placing rubber gaskets under the recoil lug is, in this case, out of the question.

Glass Bedding the Barrel and Action

The process of glass bedding is more complicated than free floating and requires more preparatory work, but the results are well worth the additional effort. In order to complete this task, you will need some materials:

a. Carving tools, X-Acto set or a small wood chisel is helpful. A good scraping tool can be made by shapening the end of a piece of half inch steel EMT conduit.
b. Coarse grade sandpaper may be used.
c. A paintbrush with about a 2 to 2 1/2 inch width.
d. Common floor-type paste or liquid wax.
e. Masking tape.
f. Bedding compound.
g. Epoxy floc (if needed).
h. Powdered metal (optional).
i. Modeling clay.

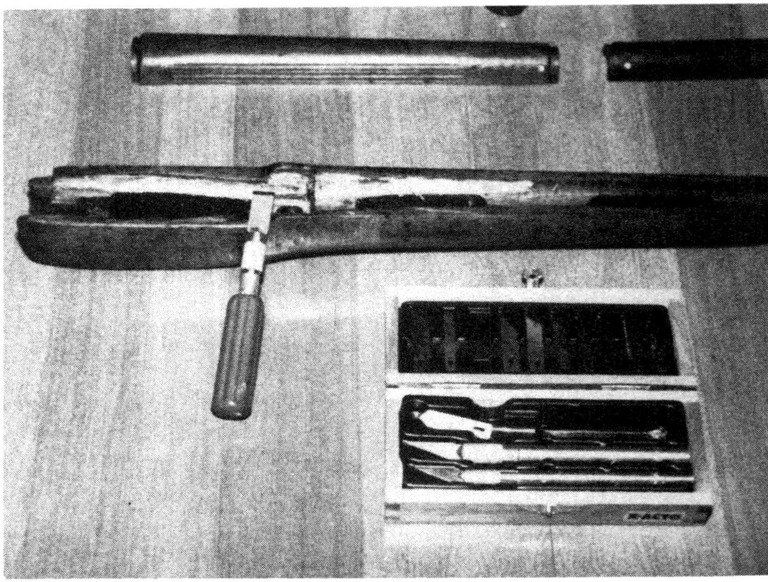

An X-Acto set is very practical for removing wood from the inlet channel. The surface does not have to be perfectly smooth as the roughened wood will provide a better bonding surface for the bedding material.

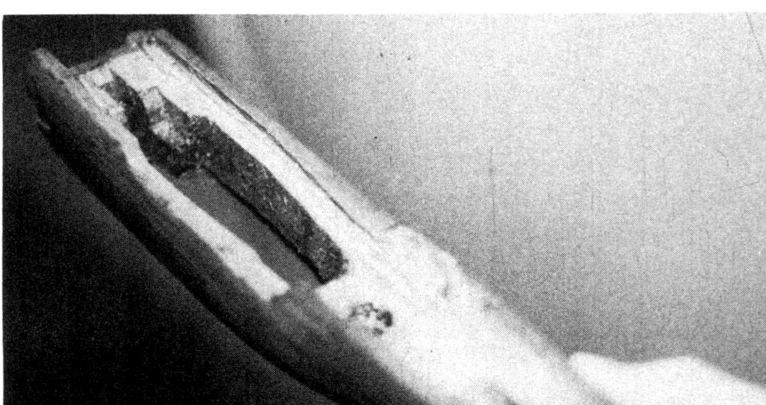

Notice the rough cut made in removing wood where the action will fit.

Preparing the Stock

The bedding job is begun by removing the barrelled action from the stock. Set the barrelled action aside for now, as the stock is going to take up all of your time for a while. What we need to do at this point is to remove wood from the inletting channel, or the entire area in which the barrel and action of the rifle rests when the rifle is assembled. There is good news and bad news about this. The good news is that it is near impossible to mess up the wood removal. If you remove too much wood, the extra bedding compound will (usually) simply fill it in. If the wood removal looks ragged and rough, great! A rough surface is a benefit, because the epoxy of the bedding compound adheres best to a rough surface. This is the reason that fine sandpaper is never used in the glass bedding process. The bad news is that unlike the free floating mentioned earlier, there is no short-cut possible here. The wood has got to come out. The best way to do this is to cut away wood by enlarging the inletted recess below the upper edge of the stock. This is the method preferred by Jim Carmichael, and it the one that looks the best when the bedding process has been completed. This procedure provides plenty of space to be filled with bedding compound, while making it look on the outside as though the stock has never been altered.

You will need to remove about 1/8 inch to 3/16 inch of wood; about 3/16 inch under the action and around the recoil lug, and about 1/8 inch along the sides of the action, with at least 1/8 inch around the barrel. When the bedding compound is in place and the stock is bolted back up to the barrelled action, excess compound has to have somewhere to go. In the barrel channel, the excess epoxy will be squished up around the barrel, which can be trimmed before it sets up completely. The recoil lug area and the area under the action can best be prepared for allowing the escape of excess epoxy by cutting a small groove across the floor of the stock (the area just behind the recoil lug) back to the magazine well. This will allow the excess material to escape into the magazine well, where it can be trimmed off later.

Two important things to remember when preparing the stock: First, cut away a portion of the stock behind the recoil lug so that quite a bit of bedding compound will fill in behind the recoil lug of the rifle. The sides of the recoil lug cut should be removed also so that there is bedding support all the way around the recoil lug, but cutting away the wood behind the lug area is frequently missed by home gunsmiths.

If your rifle has upper handguards (wood pieces which cover the top of the barrel), these will not have to be glass bedded. Handguards are used only to allow the rifleman to hold the rifle by the barrel after extended firing without burning himself. Handguards should already be essentially free floating.

The second thing to remember is when cutting away wood under the receiver, leave some supporting edges on both sides of the inletting to support the action at its appropriate height in the stock. If these ridges are not left in place, the barrelled action will be pulled too far down into the bedding epoxy.

When this work has been completed, the next step is to tape the outside of the stock with masking tape to protect it from the epoxy resin. After taping, set the stock aside and proceed to the barrelled action.

Prepare the barreled action using the paintbrush and the liquid floor wax (or paste wax and your fingers),

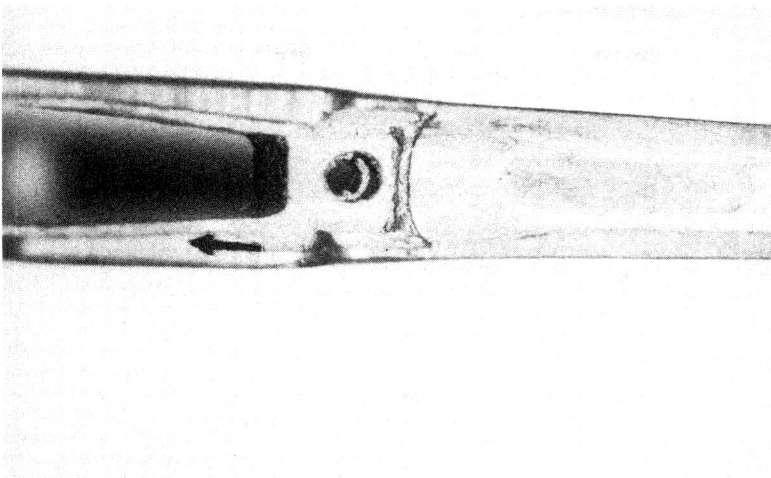

Leaving some supporting edge on both sides (arrow) will keep the barreled action from being pulled down too far into the bedding material. The round hole is the SMLE's version of a recoil lug inlet. For most other rifles, the recoil lug is a rectangular piece of steel on the bottom front of the action which seats into a rectangular hole.

paint the entire outside of the barrel and action. This will act as a mold release, permitting you to remove the barrelled action from the stock after the epoxy sets up. Make SURE you cover every inch of the steel. If you miss a spot, you have got BIG trouble. The epoxy will permanently bond to the stock AND the steel, and you will not be able to get them apart. I have heard stories of people who have done this and have had to literally chop the stock away from the barrelled action. After you have painted the entire outside of the barrel and action, set it aside to dry. While it is drying, dip the stock bolts in the floor wax and set them aside on waxed paper to dry also. Don't neglect dipping the stock bolts. If a little epoxy gets on the threads of the stock bolts, they will be permanently attached when you screw them into the action.

One thing to note is that floor wax has been largely replaced in grocery stores by acrylic floor preparations. Don't use these in the place of wax. I've never tried them out and don't know if they would work as a mold release or not. They might, but then again they might bond to the metal and epoxy both, in which case you have a real mess on your hands. I know that floor wax works exceedingly well, so I would stick with what has been used and is known to work.

After the wax has dried thoroughly, examine the action and barrel for any holes (such as screw holes). Don't worry about the holes that the stock bolts go through, but if there are any other holes, fill them with modelling clay. If there are any very deep scratches, it might be a good idea to fill these in also. What we are trying to do here is to eliminate the possibility of the epoxy locking the stock to the barrel or action. If there is an open hole in the action or barrel, the epoxy will flow into the hole and bond to the stock, and you will be unable to separate the two. If holes are filled with clay, there will be no problem.

Preparation of the Bedding Compound

The preparation of bedding compound depends upon whether the user is going to use a commercial bedding compound specifically designed for glass bedding, or whether he is going to use some other epoxy compound.

If commercial preparations made specifically for glass bedding are used, there is really very little preparation necessary, other than following the instructions that come with the compound. The only thing to be careful about is mixing the two resin parts in precisely the right proportions. There are several of these mixes on the market. "Steel Bed" contains powdered stainless steel, and provides an exceptionally strong matrix. "Acraglas" is another, a nylon based epoxy system without the addition of powdered metal. Brownells' "Glasbed" is probably the easiest to use. It comes in a plastic pouch with a divider between the two epoxy resin compartments. To use, you just pull the divider, knead the pouch for four minutes, cut one corner of the pouch with a pair of scissors and squeeze the compound into your stock recesses. All of the above bedding systems are available from Brownells at prices ranging from about $10 to $15.

Although the commercial glass bed preparations are the best, it is not necessary to use commercial preparations to glass bed a stock. Regular epoxy can be used if desired, but there will be more work involved. In this case any epoxy used should have at least a 20 minute cure time. The 5 minute stuff will set up before you have time to get it mixed, distributed in the inletting channel of the stock and get the rifle reassembled. Trust me on this. Even 20 minutes doesn't leave you as much time as you think. Don't even try to use this fast setting stuff.

If you want to use a different source of epoxy than the commercial glass bedding preparations, you will need some floc. This is a fiber which is usually, but not always, fiberglass. You should be able to get a small amount of floc from a automobile body shop, and if you can't get any from this source, you can always get some through an auto supply store, but will probably have to buy a larger amount. The addition of fiberglass floc to the epoxy will result in strength increases of about 10%; nylon based epoxies such as Acraglas are as strong as the fiberglass/epoxy matrix without the need for using fiberglass floc. When you are ready to bed the stock, mix your epoxy first, then add in enough floc (with constant mixing) to make a mixture with about the consistency of thick grease.

The strength of the epoxy matrix can also be increased by up to 10% with the addition of powdered metal. There are several types of powdered metals available, including powdered steel, stainless steel, aluminum and bronze. Don't use powdered steel. It is liable to produce rust flakes on the outside of the glass matrix, and you won't be able to do much about it. I would probably avoid aluminum also. Stainless Steel is acceptable, as is bronze. Both produce extremely strong matrices, although the bronze makes the hardened compound easier to work. If you want to use powdered metal, mix your epoxy first then add 1 part metal to 4 parts mixed epoxy. The powdered metal is available from Brownells.

If you are using white or clear epoxy, you may want to color your bedding compound to match the stock. The best way to do this is to purchase artist's dry pigment from a hobby shop or an art supply house, and add it to the epoxy as it is being mixed. Use only tiny amounts, because a little of this stuff goes a long way.

Putting the Parts Together

When you have mixed the epoxy compound of your choice, spread it throughout the inletting channel where wood has been removed. Reassemble the barrelled action to the stock, install stock bolts and pull them up snug, pulling the barrelled action down into the bedding compound in the inletting channel. When the bedding compound has cured to a kind of rubbery consistency, trim excess amounts, which have been forced out of the stock, with a razor blade or an X-Acto knife. Trim the material right up to the metal. If this is done right, it will be hard to tell that the action has been glass bedded at all.

When the epoxy has cured to its final hardness, unscrew the stock bolts about half way, until they protrude well above the level of the stock. Give the heads of the bolts a solid whack with a WOODEN mallet. It may also be necessary to whack the bottom of the barrel where it extends beyond the stock. Once this has been done, the action and barrel should come loose from the stock. If it does not, you may have missed a

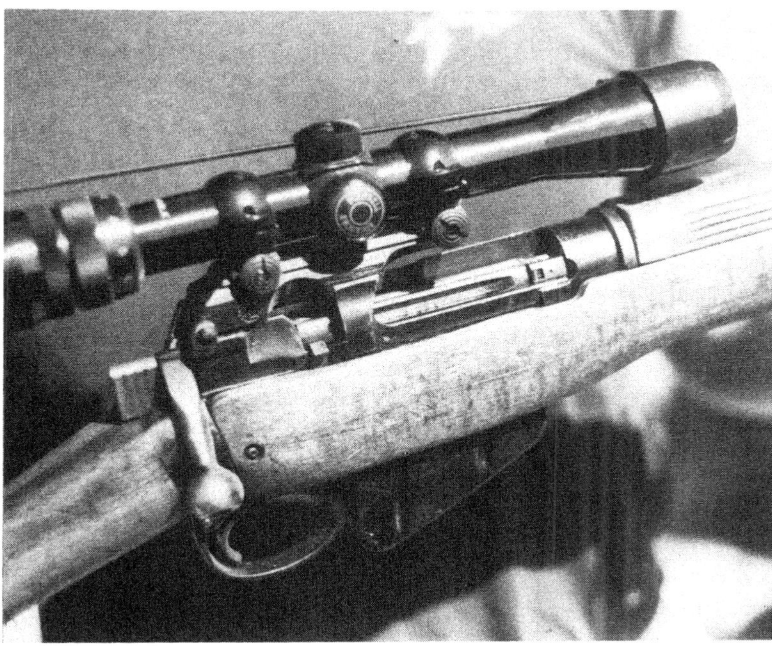

Note that with proper trimming of the bedding material around the barrel and action, there is little to no evidence of the action being glass bedded.

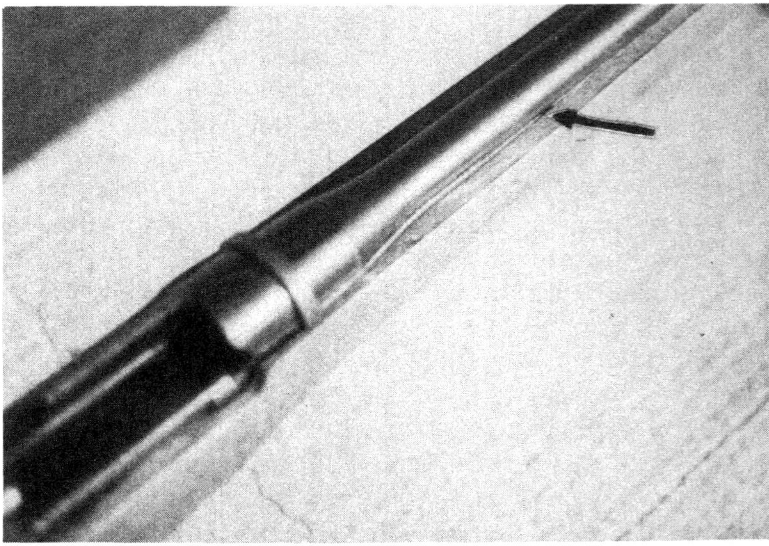

The skin-tight fit between the barrel and stock is evident. An occasional air hole can occur but it will not affect accuracy.

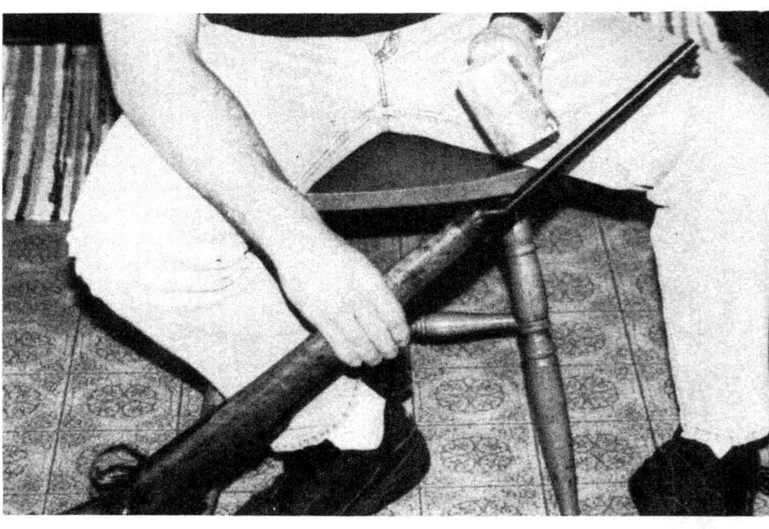

After the bedding compound has setup, a smart rap to the bottom of the barrel and action should cause it to loosen and fall out.

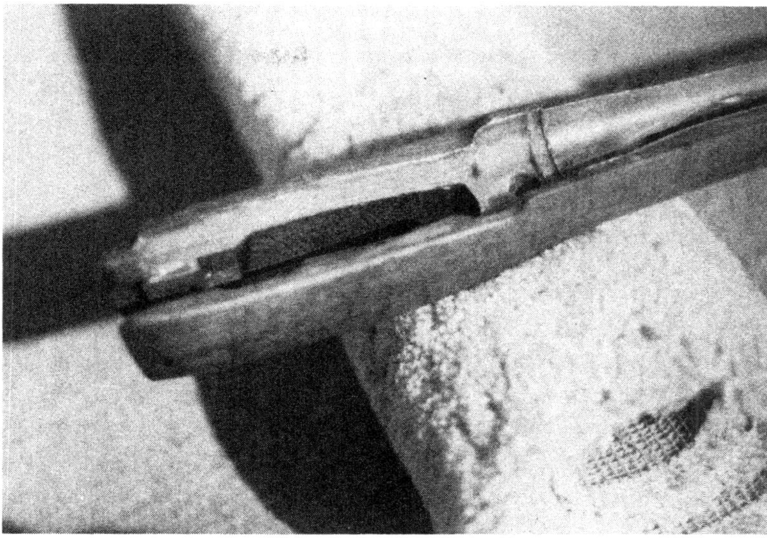

Note, in the finished glassed SMLE stock that the most critical areas for glassing are the recoil lug hole and the barrel for about the first two inches from where the barrel screws into the action, foreward.

spot somewhere when painting the barrel and action with the paste floor wax. To find out, put your rifle in a freezer for a couple of hours. The metal of the rifle contracts in the freezer, but the glass bedding does not. After a couple of hours, you should be able to remove the rifle from the freezer and the barrel and action should fall right out of the stock into your hands. If it doesn't, make sure all stock bolts have been removed, then try the mallet again.

If you really have failed to cover all of the metal of the rifle with wax, and the action and barrel have bonded to the stock through the epoxy, there is probably only one relatively easy way to get it off. You will have to heat the barrel and action up with a propane torch. Epoxy is a type of plastic, and when the metal gets to between 300 and 400 degrees, it should release, but the heat will ruin the stock in the process. If I bonded my rifle and stock together accidentally, I would probably think seriously about using it AS IS, and hope that no one asks me to take it apart for them. I would already feel like an idiot, I would hate to look like one too. As long as it is not necessary to take the rifle apart, everything should work great.

One other point: when the action is removed from the bedded stock, you will probably find holes in the bedding compound where not enough epoxy was placed in the channel at a specific spot. Some fill these holes in with a second go-round of epoxy. I recommend against it. The holes don't affect the accuracy in any way and filling them usually doesn't work very well anyhow.

Section IV
The Sniper (or Match Shooter) Trigger

Once the barrelled action has been either fitted with a synthetic stock (the expensive way to go which could conceivably require glass bedding anyhow) or the rifle's wooden stock has been appropriately glass-bedded, free-floated, or some combination of the two, it is necessary to turn one's attention to the next vital part of the rifle for modification, i.e. the trigger. Military trigger pulls vary from rifle to rifle, generally running from 5 to 8 pounds. Sometimes the pull is greater than 8 pounds and occasionally less than 5, but not often less. The best trigger pull for a sniper rifle is between 2 and 4 pounds. Four pounds is a little too much for me, but some shooters prefer a slightly heavier pull. In my opinion, the trigger pull should be no more than 3 pounds and no less than 2 1/2 pounds. Some match shooters like a trigger pull which is a true hair trigger, one that the slightest touch will release, but in my opinion when you get to trigger pulls below 2 pounds, you are in great danger of firing the rifle accidentally. In a hunting situation this could result in a tragedy involving an accidental shooting, and in a sniper situation it could result in giving your position away prematurely, an incident which could end in tragedy of a different kind.

The military trigger pull is usually a two stage affair in which the trigger finger feels resistance which must be overcome, allowing the trigger to be pulled back until a second heavier resistance is felt. When this second stage resistance is overcome, the rifle fires. The heavier the trigger pull, the more likely one's shots are going to be pulled off to one side. Remember, it only takes a tiny deflection at the muzzle to result in a large deflection at a hundred yards. and a huge deflection at 300 or 400 yards. Some shooters radically dislike the two stage military pull, and see it as a barrier to accuracy. In truth, it is only a barrier to accuracy if one allows it to become such a nuisance that it affects his shooting. Many military shooters have used the two stage pull through the years without a problem of any kind. It is more likely to be a problem with match shooters or varmint hunters who have availed themselves of commercial rifles equipped with single stage triggers, the type of trigger generally favored by this group of shooters. If the shooter allows the two stage pull to affect his shooting, he should do something to replace it. If it is not a problem to the shooter, then the two stage pull can be retained. This is primarily a matter of what you are comfortable with.

Assuming we are describing the two stage pull of a military rifle, the first resistance is overcome and the trigger is pulled back until the second (stronger) resistance is felt. At this point, pressure on the trigger mounts as the trigger is squeezed until the cocking piece is released and the rifle fires. If the trigger continues to move under the second stage pressure prior to the release of the cocking piece, this additional movement is called creep. Creep is undesirable. Ideally, there should be no discernable movement of the trigger at the second stage before firing takes place. If there is no discernable movement prior to firing, the shooter cannot anticipate when the rifle is going to fire, thus reducing the chances of pulling the shot off to one side as a result of flinching. On the other hand, if the creep is very long, it is also virtually impossible to tell when the rifle is going to release, thus the pulling of shots due to flinching should be reduced in this case also, but it is generally accepted that little to no creep in a trigger is preferable to a very long creep. When there is no discernable movement, the trigger is said to "break clean", or be "crisp".

Once the rifle has fired, if there is additional backward play in the trigger, this is called backlash. Some prefer to adjust backlash, reducing or removing it entirely, but once again this is largely a matter of preference.

As you can see, the preferred trigger in a highly accurate rifle is one in which the trigger pull is very light and in which there is little to no creep. This has been dealt with in varying ways in the past. The British, along with many other Europeans, found a solution to their trigger pull problems through the use of the double-set trigger. This is a system in which the rifle has two triggers. Only the front trigger fires the rifle. If the front trigger alone is squeezed, it fires the rifle at its normal trigger pull. However, if the back trigger is pulled first until it "sets" the front trigger (i.e. until it clicks), at this point the front trigger will fire the rifle with the lightest touch. Although the double-set trigger has been a popular approach for the Europeans, it has never caught on here in the U.S., except in old fashioned, black powder replicas.

Another approach which has found some popularity in America has been the single set trigger. This system uses a single trigger which fires at the normal pull, and is set by pushing the rear of the trigger forward. Canjar makes one of the best single set triggers in the industry (Canjar Co., M.H., 500 East 45th Ave., Denver, Colorado, 80216. 303-295-2638) if you have an interest in this type of trigger, but it is made for a limited number of rifles, primarily commercial models.

The third approach is the use of a standard adjustable trigger. These function the same as a normal trigger (there is no "setting" function to this type of trigger. You just squeeze it.), except pull weight is adjustable from about 2 to 4 pounds, and creep is adjustable, as is backlash. This type of trigger is preferable because of its simplicity and because price can be very competitive. Timney makes an excellent adjustable trigger which retails for only about $36 through Brownells (address previously given. Ask for their catalog.). There are other adjustable triggers available on the market as well, such as the Kepplinger, but these will be more expensive, and are, by and large, available for only one military rifle, the 98 Mauser. The Timney is made for several military rifles based on the Mauser design, to include the pattern 14 Enfield, the pattern 17 Enfield, the 1903 Springfield, and a variety of Mauser models to include the 93, 94, 95, 96, and 98 Mausers, as well as a variety of commercial models. Adjustments to the trigger are easily made by simply turning machine screw nuts, and in most cases the trigger assembly simply bolts up to the rifle with no gunsmithing needed. I strongly recommend the Timney trigger if one is available for your military rifle, because they are of a high quality, simple to install and easy to adjust.

The Timney trigger comes in three models, the Sportsman, the Featherweight, and the Featherweight

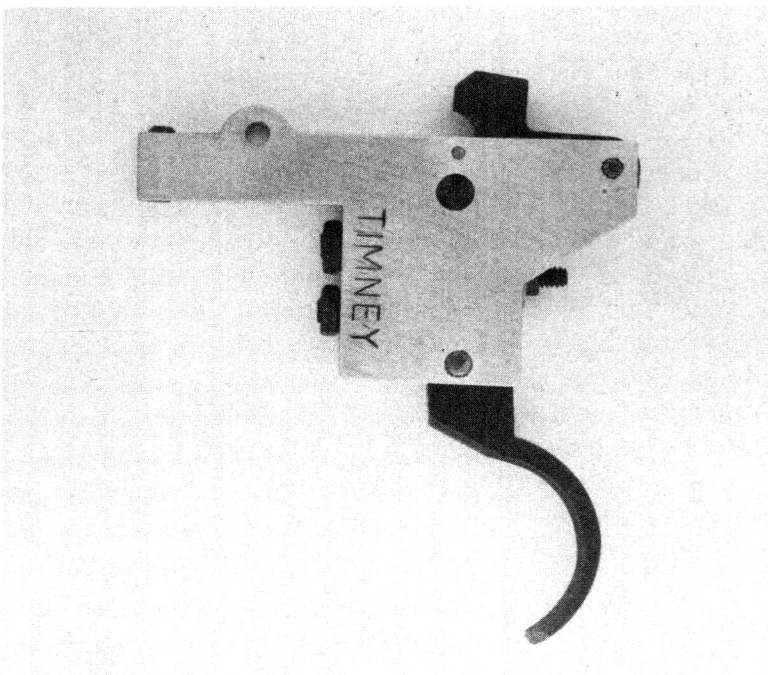

A Timney trigger for a small ring Mauser, one of many fine triggers assemblies for vintage military rifles. Photo courtesy Timney Mfg. Inc.

Deluxe, retail pricing for the three through Brownells comes in at $36.75, $55 and $67.50 respectively. The Sportsman has an adjustment range for pull weight from 2 to 4 pounds; both Featherweight models adjust from 1 to 3 pounds and the Featherweight deluxe has a safety built into the trigger assembly. As I mentioned earlier, I feel that a trigger pull of less that 2 pounds is not advisable for a sniper rifle. I recommend the Sportsman for the Enfields since these can effectively use their own stock military safeties, but recommend the Featherweight Deluxe for all Mausers and Springfields unless the shooter is planning on mounting a low clearance safety or a side swing safety on their rifle (more about this in the next section) . The Mauser type safety is inadequate when telescopic sights are mounted because there is a problem with clearance between the safety and the telescope tube. There is usually not enough room to engage a Mauser type safety with a telescopic sight in place. With the Featherweight Deluxe, the safety is on the trigger rather than on the rear of the cocking piece, in which case engaging the safety with a telescopic sight in place is no longer problematic.

With some rifles, such as the SMLEs or the Moisin Nagants, there are no adjustable trigger assemblies available and the shooter is forced to do a little gunsmithing to bring his trigger pull into line with what is required for a sniper rifle. Incidentally, this same general procedure can be used to reduce the trigger pull of any military rifle, but I feel that the adjustable trigger is a better way to go if your rifle is one that has a Timney trigger available for it, because while the following manipulations will reduce trigger pull, it will do nothing for creep or backlash. Gunsmithing your trigger pull, however, is the cheapest way to go, unless you own no polishing tools or supplies at all.

I do not recommend trying to remove the double stage pull from the stock military trigger. I have seen descriptions of how the first stage in a military trigger pull can be removed through the use of set screws or various mechanical set-ups. What these typically do is to mechanically force the trigger into a semi-pulled position to get rid of the first stage. Doing this makes inadvertent firing of the rifle likely when the bolt is closed. This is about as risky as having a rifle with a truly "hair" trigger. Don't do it. Gun Parts Corp., West Hurley, New York, 12491, has replacement single stage triggers for 98 Mausers and Springfields for about $5, but these only eliminate the two stage pull. They are not adjustable in any sense.

In order to reduce your military trigger pull, you are going to have to disassemble your rifle and understand how the trigger engages the sear, which in turn holds the rifle at full-cock. The use of one of the disassembly texts described earlier which contains exploded diagrams would be helpful here. In the case of the SMLE, the trigger bears on the bottom leg of a spring loaded "L" shaped piece of metal, which is pinned to the action at the intersection of the legs. This piece of metal is the sear. The top leg simply snaps into a slot machined into the rear of the cocking piece. As the trigger is squeezed, it bears down on the bottom leg of the sear, forcing the top leg out of the slot in the cocking piece, allowing the rifle to fire. It is as simple as that.

Gunsmithing to Reduce Trigger Pull

Disassemble your rifle. Remove only what you need to remove to get the sear and the cocking piece out to where you can handle them. If your sear is pinned and the pin won't come out, don't try to remove it, but work with it in place on the action. Pins come in two varieties: easy to remove and impossible. If a few taps of a punch won't release the pin, it may be locked in place. This is done by drilling into the installed pin at an angle and then driving a slightly oversized pin into the hole, or peening the head of the pin to where it is oversize relative to the hole it is used in. In either of these cases a gunsmith would have to drill the pin out to remove it. You don't want to do that. In such a case, just work with the sear in place. It is inconvenient, but not impossible.

Examine the sear for burrs (i.e. rough, often ragged pieces of metal on the edges of the parts). If you find ANY burrs, remove them with a small file. Make sure to remove only the burrs. After removing burrs (if any are present), use a fine stone used for sharpening knives such as an Arkansas stone, or even better, a gunsmith's India stone, and stone the filed surfaces smooth. Using crocus cloth (obtainable at any hardware store), burnish the surfaces on the sear and the sear slot on the cocking piece until they are as smooth as glass. The idea is to polish these pieces where they touch, reducing friction. Reassemble the parts. On all bearing surfaces (surfaces where metal touches metal) place a small amount of high quality gun grease. For this, I recommend using Brownells' Action Lube Plus or Chip McCormick's Trigger Slick, a grease which not only lubricates but polishes surfaces making the pull even smoother with use. Both are available through Brownells. If you do not have access to either of these, a good grade of milspec moly based gun grease will also do nicely.

On many guns, the simple procedure just described

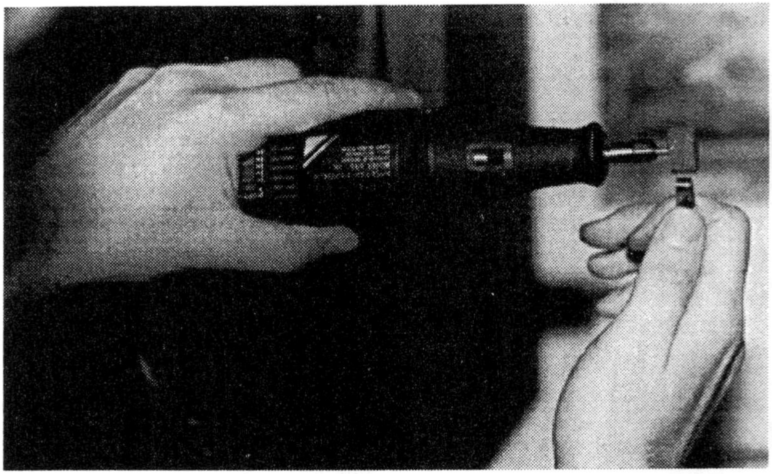

Thinning metal on the sear spring to reduce trigger pull. Only the thickest parts of the spring should be thinnned. Removal of too much metal may cause the spring to fail. Bright metal can be blued using a good cold blueing product.

along with lubrication is all that is needed to produce a perfectly usable trigger pull. Reassemble the rifle and try the trigger. Look for a light easy pull that feels right to the shooter. A better method would be to measure the pull weight, but this requires the use of a trigger pull scale. If you have one of these, look for a trigger pull between 2 and 4 pounds. If the trigger pull either feels or measures too heavy, you will have to go to the next stage of modification.

Look at the trigger spring. You will see that it is very thin on the ends and quite thick towards the middle. Chuck a small cylindrical grindstone in a Dremel moto tool, and carefully grind down the spring in the thick areas only to reduce spring thickness (you can also use a file to do this, but it is more time consuming). You have to be very careful not to grind on the thin ends of the spring and you can't remove too much metal. If you do, the spring will break under use. But you should be able to reduce the thicker part of the spring by about 20% or so with no problems.

After grinding the spring down, you may want to reblue it before re-installation. Use a touch up blue, or Brownell's Oxypho Blue to do this. Slide the spring back into place, and try the trigger pull.

In the test SMLE as it was received from the wholesaler, I found that the first stage of trigger pull broke over at 58 ounces, and the second stage fired the rifle at 6 pounds of pressure even. Not bad for a military rifle. After using the above techniques, the first stage broke over at 46 ounces and the second stage fired the rifle at a clean 3 pounds 8 ounces. This is slightly heavy for my preferences, but is certainly within the range of a match trigger.

Some occasionally try to alter trigger pull by changing the angle of the sear slot, where the sear engages the cocking piece. Don't ever do this. If you want to smooth up the slot, use a fine stone (never a file) and/or crocus cloth, but take care to retain the original angle. Messing around with the angle of sear engagement will produce a rifle that will fire when it is just bumped. Don't do it!

CAUTION: IF YOU ARE GOING TO REDUCE TRIGGER PULL ON YOUR RIFLE BY ANY MEANS, GET IN THE HABIT OF LEAVING YOUR FINGER OUTSIDE OF THE TRIGGER GUARD UNTIL YOU ARE READY TO FIRE. EVEN A CASUAL BRUSH OF THE FINGER ON THE TRIGGER WILL PROBABLY CAUSE THE RIFLE TO DISCHARGE. In truth, this is a safety procedure that the shooter should utilize anyhow, but it is especially critical after reducing the trigger pull.

Section V
Sights for the Sniper Rifle

This is perhaps the most critical part of the whole conversion process. You can have the most expensive, innately accurate rifle in the world, but if you have a crappy sight system, the accuracy is lost and the rifle is worthless.

At the present time there are basically six types of sighting systems in use worldwide. The first is called "metallic sights" or "open iron sights". These are the old V notch sights used by most military rifles in times past, but are no longer found in the more modern military rifles; they are made up of a V notch rear sight and a post front. The V notch is simply a V shape cut in the rear sight leaf. One aligns the front sight post into the V notch (some use a square notch) then sets the aligned sights on the target and fires. It sounds simple, but there is a big problem with this system. In order to make it work, you have to focus your eyesight on three planes, i.e. the rear sight, the front sight, and the target. Unfortunately, the eye wasn't designed to focus on more than two planes at the same time. What happens is this: You focus on the rear sight and front sight, bringing them into alignment. While you are doing this, the target goes out of focus. Once you have front and rear sights aligned, you bring the sights to bear on the target, usually keeping the target and front sight in focus. At this point, the rear sight winks out into oblivion, and although you now have the front sight and the target in focus, you don't know whether the front sight and the rear sight are in alignment any more. So you change your focus to the front and rear sights again, at which time the target vanishes. So you end up playing this juggling game between sights and target until you eventually squeeze off the round. This system is the oldest of all sight systems, and can work reasonably well at short distances if you fire the rifle from a rest position (especially at a stationary target), and if you are younger than 40 yrs. old. The older you get, the less able you are to change eye focus rapidly (this is why we codgers need bifocals). If you are an older person, taking a snap shot at a rapidly moving target, using open iron sights, you have the proverbial "snowball's chance in hell" of hitting your target. If your rifle is equipped with open iron sights, I would recommend replacing them, even if you aren't doing a sniper conversion. Open iron sights are the least effective of all the sight systems. They came from the distant past, and are better off left there. This type of sight could be termed a low technology (or "low tech") development, as it requires neither optics nor electronics.

The second sight system, also a low tech development, is known as "peep" sights. These are sights comprised of a disc with a hole in it for the rear sight and the normal post front sight. The rear peep is usually adjustable for both windage and elevation, unlike many of the metallic sights. The oldest iron sights were not adjustable for either windage or elevation, but were fixed. They were adjusted on a one-time basis for the original owner by filing down the front sight to provide the proper elevation. Windage adjustments (if required) were done by "Kentucky windage", or basically guessing how much to hold left or right to compensate for wind or whatever. More modern military rifles, such as the Mauser, had metallic sights which were adjustable for elevation, but not windage.

When using peep sights, the problem of eye focus in several planes is eliminated. You just look through the peep hole, set the front post on the target and squeeze off the round. You keep the front sight in focus and the target in focus, and forget the rear peep. It is not necessary to try to center the front post in the peep hole; your eye does that automatically. You don't have to even think about it. Most of the military rifles made at this time are equipped with peep sights. The peep sight is far superior to open iron sights. If you want a good, reasonably accurate military rifle, but are not interested in going to the trouble of doing a complete sniper conversion, take the V notch sights off and install a peep sight. More about peep sights a little later.

The third system available is that of the electronic floating dot, such as the Aimpoint or Mirage. These systems look like somewhat oversized telescopic sights, although most do not provide any magnification (some provide slight magnification). These systems electronically produce a red dot that seems to float in the air as you look through the tube. In this case, one only has to focus on only a single plane, i.e. the target. You simply put the dot on the target and squeeze the trigger. This is a high tech system in that it utilizes both optics and electronics, and it does offer some nice benefits, but has significant drawbacks as well. Benefits are that the sight is quick to bring on target, it is incredibly simple to use and is quite accurate at short ranges. Drawbacks are that some units are bulky, they are dependent on batteries, they are extremely expensive and they are strictly a short range system. Because of the drawbacks, these are not considered to be suitable for sniper use.

The fourth system, another high tech addition, is the

laser sight. These are small lasers attached to rifles, which project a small dot onto the target. Lasers are gaining in popularity for reasons that I confess not to understand, although they are quite easy to use. You just move the barrel of the rifle in much the same way you would move a flashlight, and when the laser dot touches the target, you squeeze off your shot. The laser creates a very tight dot out to about 200 yards., but beyond this distance, the dots begin to open up until they get so large they become unusable. In his book The Terrifying Three (1989) Duncan Long also points out some additional drawbacks of the laser. First, it is not usable in other than dusk or night applications. Bright daylight washes the beam out making it impossible to see. Also, in conditions of smoke or fog, the laser does not generate a dot, but a highly visible beam which can easily be traced back to its source. Even without smoke or fog, if one is at the right angle in front of the beam, the location of the device can be seen when the laser is used. For this reason, Long recommends that if the shooter plans on staying in one piece he should change positions between each activation of the laser. Prices of these devices vary from a low of $99 through the $300 range, depending upon the quality and features.

The fifth sight system (another high tech development) is that of the night-vision scopes. These are the specialty scopes used strictly for night use. They are based on military technology, and utilize a light cascade effect to permit the shooter to see with near daylight clarity using only starlight. Although these definitely have a place in long range sniping in darkness, they are not going to be dealt with here, because the point of this manual is to allow the shooter to add a highly accurate sniping system to his gunrack at a reasonable price. These scopes often cost several thousand dollars. Since the fall of the Soviet Union, quite a large amount of starlight Soviet equipment has been dumped on the American commercial market at prices much better than what one would pay for comparable American equipment, but it is still up in the thousands of dollars, and thus I doubt that the average middle-class American could afford it. If he could, he doesn't need this manual.

The last sight system is the standard telescopic sight. I have termed this system as mid-tech because it does utilize optics, but not electronics. There may be some carryover between types, however, as some of the more modern telescopic sights in use today are being manufactured with a small illuminated dot at the point where the crosshairs meet (for use in low light conditions). The illuminated dot necessarily involves the use of some electronics and requires a battery to maintain illumination, but the dot is not necessarily needed and the scope can be used without illumination, relying only on the crosshairs and the optics of the scope itself.

Telescopic sights have been traditionally used by sniper rifles for many decades and to this day is probably the most effective sight system for the money. Its positive attributes include the fact that magnification compensates (up to a point) for deteriorating eyesight in older shooters, it requires focus on only a single visual plane, it allows one to distinguish targets at distances which would be invisible without the telescopic optics, and although they can be quite expensive (depending upon models and manufacturers), the system has been refined and mass produced over the years to where cost can be quite reasonable if one looks in the right places.

In summary, there are six sight systems in use at the present time: two low tech systems, of which only peep sights should be used by serious shooters; three high tech systems which are not suitable for the use given in this manual due (in part) to exorbitant expense, range limitations or lack of effectiveness in full daylight; one mid-tech system which lends itself well to the sniper conversions. It is easy to see that practically speaking, assuming that the reader has a limited budget to draw from, there are really only two practical sight systems for use, namely the peep sight and the telescopic sight. The peep sight is a fine sight, rugged and likely to outlive the telescopic sight by a considerable margin, but limited for the most part to applications other than sniper. Although the telescopic sight is more fragile than peep sights, it is also capable of much finer accuracy, as anyone who has spent a great deal of time using both will tell you. As with most things in life, there is a certain amount of trade-off between desirable and undesirable characteristics when trying to arrive at a suitable system for sniping. Ultimately, although the search for a better sight system continues, the telescopic sight has emerged as the clear winner. Accordingly, the remainder of this section will deal briefly with peep sight installation (briefly because the sniper rifle will not be equipped with a peep sight) but will emphasize the selection, purchase, and installation of the telescopic sight, and any attending modifications to the rifle which might be required.

Peep Sights: Selection and Installation

The sniper rifle will not employ peep sights, but any rifle with metallic sights will have its accuracy improved by the addition of peep sights. There are a variety of manufacturers of peep sights, producing a wide range of sights, which can range in price from around $30 to well over a hundred. There are generally two types of peep sights when it comes to mounting styles: tang mount sights and receiver mount sights.

Tang mount sights attach to the receiver tang, or the metal extension at the rear of the rifle through which the stock is attached to the receiver. The type of rifles typically using the tang sight are older ones, such as the model 99 Savage, many of the lever action Marlins, the lever action Winchesters in models 92, 94, 95, and 1886, as well as the old falling block rifles such as the Browning High Wall, the Sharps, and the rolling block Remingtons. Many of these rifles are already drilled and tapped for tang mount peep sights. One only has to align the base of the peep sight with the tapped holes on the tang and install the machine screws.

Receiver sights are more common. Some military surplus bolt-action rifles come already equipped with adjustable peep sights, such as the 1903A3 Springfield, the pattern 14 and pattern 17 Enfields and the number 4 SMLEs. However, many military rifles, including essentially all bolt action Mausers, come equipped with the old metallic sights. Many commercial rifles are already drilled and tapped for receiver sights even though they come with metallic sights, because metallics are extremely cheap when compared with peep sights. I guess you really do get what you pay for. Other commercial rifles will not come equipped with peep sights, nor will they be drilled and tapped for receiver sights. If this is the case, you will have to either drill and tap the holes yourself, or have someone do it for you.

There is a large variety of receiver peep sights on the

market. Two good ones are the Lyman receiver sight and the Williams receiver sight. These can be found in the Brownell catalog at modest to moderate prices. The Williams Foolproof peepsight utilizes internal micrometer adjustments, as does the Lyman receiver sight, but the Williams 5D uses mechanical settings (aligning external marks on the sight, the moving parts being held securely in place by locking machine screws).

Williams Foolproof Peep Sight

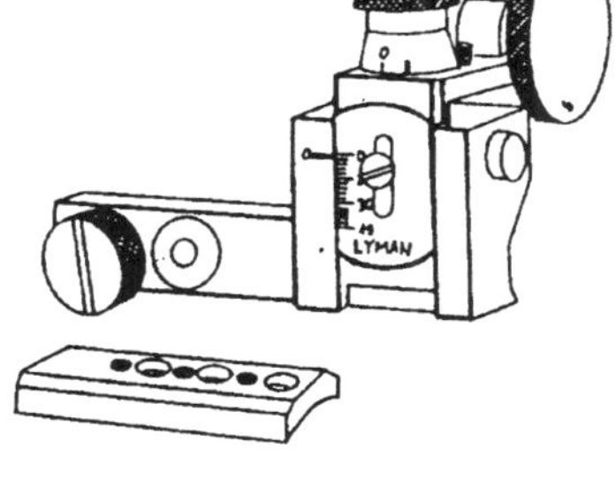

Universal peep sight by Lyman.

A Williams 5D peepsight mounted on a No. 1 Mark III SMLE

Micrometer adjusting sights are the better of the two, but are the more expensive of the two as well. The micrometer adjustments are made using a small screwdriver. Models with external knobs instead of micrometer screws are also available, and are by far the most convenient, but these are also by far the most expensive.

Selection and installation is relatively simple. Refer to a catalog such as Brownells, select the sight you want (Williams, Lyman, etc.) then refer to the chart given in the catalog under your selected sight's name. This chart will list a variety of rifles along with sight models which will fit each particular rifle. Look up your rifle, get the appropriate sight model from the chart and order it in. If you have a rifle which is more esoteric and it is not listed in the chart, the best thing would be to call the company and ask which model would be best for your rifle. If the company can't help you, you are going to have to consult a gunsmith or use your own best judgment.

Once you have received your peep sight, just follow the directions which come with the sight. If you will have to drill and tap the receiver, the next step will require three hands. Hold the disassembled sight onto the receiver at the position shown in the instructions. Insert a center punch through the holes which will be used to mount the sight on the rifle and give the punch a smart rap with a small hammer. When you lift the sight away from the receiver, you will have small indentations at the point(s) where drilling is required. Drill with drillbit size as per instructions and tap with the tap size required by the manufacturer.

Drilling should be done with a drill press (with the rifle clamped to the drill table in a level position), but I have seen a few jobs using a hand drill which were passable. Hand drills are notorious for messing up the hole, so unless the receiver metal is relatively soft and/or the drill operator is exceptionally sure of his steadiness, forget about using a hand drill and look for someone with a drill press.

Once the holes have been drilled, they need to be tapped for the machine screws that will go into them. At this point, a few things need to be mentioned about taps. These are threading tools which are used for cutting machine screw threads into steel. There are three types of taps– taper, plug and bottom taps. The taps you usually get at the hardware store are generally taper taps, or taps with a long gradual taper on the end of the tap. This type of tap is designed to tap all of the way through a piece of metal, and is especially useful in tapping holes in flat pieces of metal. The plug tap has a much shorter taper and is used to start cutting threads in blind ended holes. The bottom tap has almost no taper at all, and is designed to tap all the way to the bottom of a blind ended hole. In certain situations, you will not want to drill all the way through a piece of metal, such as when drilling a hole in a barrel. In this case, the plug tap is used to start threading in the shallow hole cut by the drill bit, then the bottom tap is used to tap to the bottom of the hole. Sometimes you will drill all the way through a piece of metal, but there will not be enough clearance on the other side of the hole to allow a taper tap to cut threads all the way through, so you may have to finish cutting threads with a plug tap. Plug and bottom taps are not always available through hardware stores, but they should always be available from machinists supply houses and will always be available from gunsmith supply houses such as Brownells.

Another little item of note is that most taps sold in hardware shops are HS taps, meaning they are made from the same steel used to made high speed drillbits. You are better off not getting these. When one breaks off in a hole you are tapping (and you might as well accept that sooner or later this is going to happen), they can be incredibly difficult to remove. If you hit one of these with a center punch, they just peen over and make the broken tap that much harder to get out. Carbon taps are not only hard, but brittle as well. These have the advantage that if one is broken off in the hole you are tapping, you can hit the tap with a center punch, the broken tap will shatter, and you can pick the broken pieces out of the threads with a pair of tweezers. This makes removing broken taps very easy, something you will appreciate after you break one HS tap off in a hole and have to remove it. Carbon taps are sold through machinists and gunsmith supply houses.

Getting back to tapping your rifle for peep sights, assuming you have kept the rifle clamped in a level position during drilling, the best way to begin the tap is to put it into the drill press chuck, lower the tap to the hole and using HAND PRESSURE ONLY turn the chuck to get the tap to start into the drilled hole. This will ensure that the tap is started in a perfectly vertical position. DO NOT TRY THIS WITH THE DRILL PRESS TURNED ON AND THE CHUCK TURNING UNDER ELECTRICAL POWER. If you do, you will simply break the tap off in the hole, then if you are using HS taps, you have got a REAL mess on your hands. Use hand power only. Once the tap has started vertically and it has cut several threads in the hole, the rifle can be removed and the rest of the hole can be threaded by hand. The first several threads will hold the tap in alignment for the remainder of the work. Lubricate the hole liberally with cutting oil, both when drilling the hole and when cutting the threads with the tap. Turn the tap in for a quarter to a half a turn, then back it out allowing the chip breakers on the tap to do their job and clean the threads. Clean the metal chips off the tap frequently. Put a little cutting oil in the hole, turn the tap back into the hole until it pulls up tight, then turn the tap another quarter to half a turn before backing it out. Repeat the process until the hole is tapped to its entire length. Go slow. Rushed jobs usually end with a broken tap.

Once the holes are drilled and tapped, clean all oil out of the holes using alcohol or some other oil solvent (such as trichloromethane), position the sight over the tapped holes, put a drop of BLUE loctite on the threads of the mounting screws and screw them firmly into place. Reassemble the rest of the sight as per instructions, and you are ready to go to the range for sighting in.

When obtaining loctite, a liquid threadlocker, which is available at all hardwares and auto supply houses, get only blue loctite. This sets up firmly, but allows you to remove the screw at a later date with little more than firm resistance to the screwdriver. If you get red loctite, the only way you will ever be able to remove the screw is by heating the parts with a torch. Don't use red loctite unless you plan on making your installation a permanent one.

The Telescopic Sight

Telescopic sights have been used on sniper rifles since the American Civil War. In those days the telescopic sights ran nearly the full length of the musket barrel, and their tubes were made of brass. They were very heavy units compared to what we have at the present time, but they were extremely effective.

The military has experimented with a wide variety of telescopic sight systems in the past, and have pretty much settled on single power scopes over the variable powered ones. At various times rangefinding devices have been incorporated into the scope, but these have generally been found to be unreliable. The military uses two man teams for sniping at the present time, a shooter and a spotter. The spotter uses rangefinding binoculars to locate targets for the shooter. The spotter usually makes very brief statements to the shooter designating target locations and ranges, such as "East window, front of the house, 600 yards.; barn door, 250 yards.," etc. With knowledge of his targets and their ranges, the sniper can set his sights as needed.

The preferred scopes used by the military at the present time are 10x Unertl scopes. These are expensive target scopes, retailing at about $230, but they are available to the public. As mentioned earlier, the USMC used 3X9 variable Redfield scopes at one time, but according to deHaas, they were discontinued, presumably because of the fact that changing the magnification results in changing the point of impact for bullets. Snipers using these scopes usually set them on the maximum magnification and locked them down, never changing power settings from that point on. I find that I still prefer the 3X9 variable in spite of what the military has done, for one very strong reason. I am no longer a young man, and my eyes need all the help they can get. The military, on the other hand, does not seem to have a shortage of young men to fill the sniper MOS.

High magnifications are extremely helpful at very long distances, but at much shorter distances, the high magnifications work against you. If you are too close to a target, you will not be able to get an adequate sight picture without reducing the magnification. In fact, if the magnification is too high, you may not be able to see anything. It is better to see what you are shooting at, even if you have to guess, to some extent, how much of a change will occur in bullet placement when the magnification is altered, than to be unable to see the target at all. If you opt for a 3X9 variable power scope, I would advise that you set it at the 8x setting and change it only when you have no choice, and then reduce power only enough to bring the target into clear focus. Setting it at 9x is also acceptable if the scope produces a clear picture at this power; some become slightly out of focus at 9x. Also, spend some time on the range determining how much of a difference in impact occurs when you go from one power to another. If you are thoroughly familiar with such information, you are better equipped to handle a situation in which reductions in magnification might become necessary.

Readers should realize one thing right from the beginning: the ideal sniper scope does not exist. Such a scope would be very small, to reduce bulk and blend in with the natural contours of the rifle. It would have a tube that was strong as steel yet light as a feather. It would provide optics with an exit pupil of greater than 6 at all power settings for maximum light gathering in low light environments. Optimally it would possess both a rangefinding device and a bullet drop compensator, neither of which would be subject to getting out of adjustment or breaking down. The ideal sniper scope would also have reticles etched in unbreakable glass optics, whose windage and elevation adjustments would be easy to make, while being positively locked at each setting, making it impossible for these to be bumped or moved out of adjustment. Sounds good, doesn't it? Unfortunately, no manufacturer has ever been able to make a scope employing all of these features, and the truth is, even if one did the expense would be so high that no one but the government would be able to afford one. This leaves us in the situation of having to do the best we can with the affordable choices we have available to us. Scope reticles (which produce the crosshair pattern) are generally made from wire, and at times they do come apart or sag under hard use. Both deHaas and Cooper have recommended the manufacture of a scope reticle which is etched in the glass of the optics, making it essentially fail-proof. The scope could be adjusted in the mount itself, rather than internally within the scope, further making it fail-proof. Few manufacturers have answered the call of these writers, although occasionally one finds scopes such as the SN-4 Omni-Purpose Tactical Format Military and Police Sighting and Ranging System (from U.S. Optics) which employs a circle dot reticle etched in the glass of the optics. Etched reticles are not common.

Although some of the characteristics of an ideal sniper scope appear to be well beyond what we have available to us at our present stage of technology, others are more realistic and more in line with matters that the shooter can (and should) pay attention to and try to obtain in any scope that he might consider purchasing. For example, the ideal sniper scope would also be hermetically sealed against moisture. Ideally the scope tube would be a single piece of metal, which makes waterproofing relatively simple, and ensures that the scope is more likely to stay that way under hard use. At the present time virtually all scopes claim to be waterproof.

The ideal sniper scope should be made of very strong material, preferably steel. Although some commercial models are made with steel tubes, most utilize aluminum to reduce weight. Aluminum alloys can be quite strong, however. DeHaas recommends tubes twice the thickness of commercial hunting rigs to eliminate the possibility of denting, bending or breaking, but this would entail both added weight and added expense, something that no manufacturer has, to my knowledge, opted to go with.

External adjusting knobs for windage and elevation are important for ideal sniping applications, less so for hunting applications. Scopes which are equipped with external adjusting knobs are also considerably more expensive, costing from $300 to $1000.

Optics should ideally be coated for maximum light gathering, but once again, all but the cheapest of the cheap come with coated optics.

A few scopes have bullet drop compensators, the Shepherd scope (Shepherd Scope Ltd., P.O. Box 189, Waterloo, NE 68069) being, at least by appearances, the best of the lot. Shepherd's bullet drop compensator incorporates trajectory correction with a kind of scaled rangefinding. This involves the use of a series of circles (having standardized diameters) on the vertical crosshair below the crosshair intersection. The crosshair intersection is zeroed at 100 yards. Each circle below the intersection represents where the bullets will hit at increases in range by 100 yard increments. In other words, at the crosshairs the bullet will hit dead on at 100 yards. At the first circle below the crosshairs, the bullet will hit dead on at 200 yards. At the second circle the bullet will be centered for 300 yards, etc., all the way up to 1000 yards. The rangefinding is done by the size of the circles. If a deer can be bracketed by any given circle (from shoulder to brisket) then he is at the right range for the shot. If he is too small to be bracketed

One on Shepherd's "Counter Sniper" models. Photo courtesy of S;hepherd Scope Ltd.

External windage and elevation knobs are important for ideal sniping scopes. If external knobs are available, adjustment can be made out to 1000 yards. If coin slots are the means of adjustment it is better to sight the rifle in at 200 to 300 yards and leave it there; making adjustments out to 500 yards use "Kentucky Windage." Photo courtesy of Bausch & Lomb, Sports Optics Divisions.

Target scopes are generally the most expensive because of quality and refinements such as the external adjusting knobs. Photo courtesy of Redfield - Denver, CO.

Shepherd's "one shot zero" process saves ammo and time. Drawing courtesy Shepherd Scope Ltd.

by the circle, one just moves down the circles one at a time until a circle is found which does fit, and he squeezes off the shot. In a sniper situation the same process is used, but a man would be (presumably) bracketed from shoulder to shoulder. Mechanical rangefinders in sniper scopes have not been successful in the past, because they generally are just one more thing to go wrong when the sniper can least afford it, but the use of scaled circles seems to be a way out of that trap, providing a reasonably accurate solution in a way that cannot break down.

Shepherd's sales brochure contains a testimony from a sergeant in the Nebraska National Guard who conducted an experiment to compare the Shepherd scope mounted on an M-14 rifle with two M-60 machine guns. They used targets ranging from 300 to 700 yards. The M-14 shooter was given 20 rounds and the M-60 machinegun shooters were provided with 100 rounds each of ammo. All began firing at the same time and all

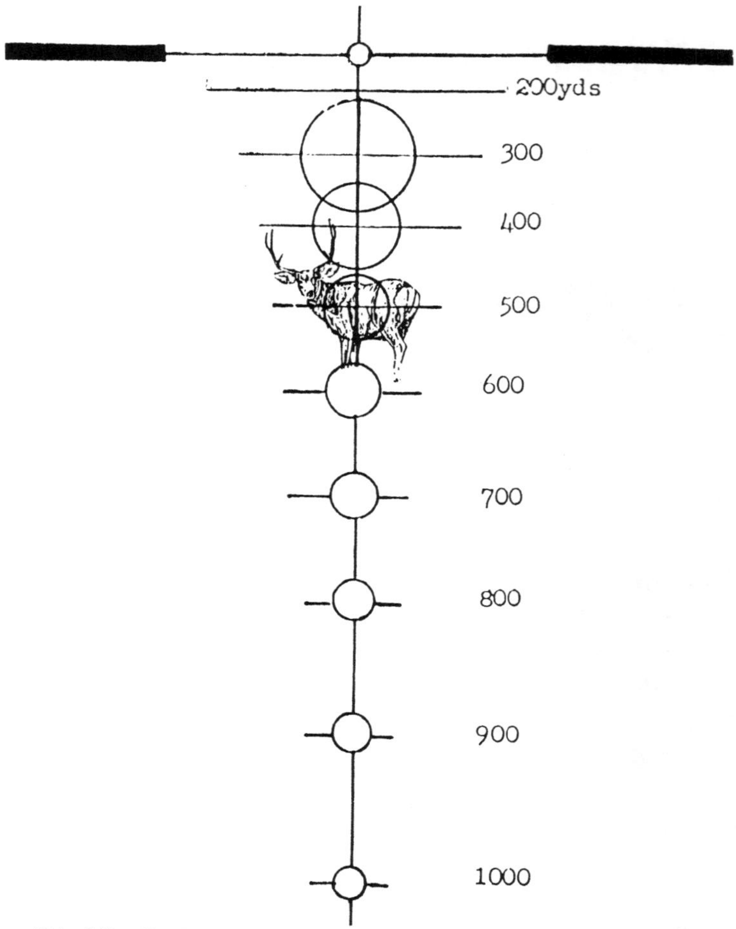

World's **FASTEST** and **EASIEST** Scope to operate.
Works even on running game.
ONE STEP INSTRUCTIONS

Put the circle on the deer that is the same size as he is—(shoulder to brisquet) and SHOOT! THAT'S IT!

Rangefinding with one step action using a Shepherd scope. Drawing courtesy of Shepherd Scope Ltd.

quit when the last M-60 ran out of ammo. When the smoke cleared the two M-60s had hit a total of 6 out of 11 targets between the two of them, using 200 rounds of ammunition and the M-14 sniper had hit 8 out of 11 targets with 12 rounds. The same sergeant claimed that at 3x settings many of their people, shooting at night, shot as well or better than others in the group using starlight scopes (the president of the company commented that if those using the Shepherd scopes would have set their magnification to 5.5x instead of 3x, they would have fared even better against the starlight scopes.)

I confess to being very impressed by what I have seen on paper where this scope is concerned. The sniping possibilities appear to be excellent, but I admit that I have never had an opportunity to test one out firsthand. I have no information, for example, on the Shepherd's ruggedness in the field. The reason can be stated in one word: cost. The various models run from $478 to $622 retail. However, I must admit that if I had the money to spend, I would probably get the Shepherd scope over all of the other competitors in the field. This is the one that seems most likely to allow one to reach that elusive 1000 yd. mark with sniper accuracy, assuming the rifle itself has such accuracy potential.

Buying a Scope on a Budget

Once we have decided to buy a scope, there are several ways to obtain one for a relatively low price. These would include the following:

1. Garage sales. My wife has twice picked up telescopic sights for less than $15 at garage sales. This is more likely a possibility in communities and states where big game hunting is common (such as Colorado).

If you see a scope which is a good brand priced at next to nothing, buy it, even if it doesn't work. One of the scopes my wife bought was a Weaver, which wouldn't zero. I sent the scope to the company, asking them to bill me for the repair. They repaired it for me free of charge. The better scopes have lifetime warranties on them, and for these, this type of service is not unusual. Don't try it on the cheaper ones, though. This kind of buy is admittedly a "fluke" that you run onto occasionally, but you don't see it every day. Thus, we may have to look elsewhere for our bargains.

2. Pawnshops. These may have scopes for sale, but they will usually be the cheaper scopes and the prices on them will be close to what you would pay for them new. The price will look cheap, but this is because the scope itself is cheap. Although it is possible to find a Weaver, Redfield, or Nikon in a pawnshop, it is unlikely. Although it doesn't hurt to look (since surprises are always a possibility) I wouldn't spend too much time fooling around in pawnshops.

3. Secondary outlets. Stores such as Wal-Mart have outlets to which they sell items that have been returned by the customer for some reason, the item is not suitable to return to the Wal-Mart shelves for sale to the public and the chain store cannot return the item to the wholesaler. We have one of these locally, and it has consistently produced scopes of various brands for $20 to $25. In stores like this, however, don't forget that these were originally RETURNED to the chain store because there was some feature about the scope that the purchaser didn't like. It MAY be broken, and secondary outlets do not allow returns. So before you buy one of these, either make sure that you can correct what seems to be wrong, or make sure that the quality

and brand are high enough that you can send it to the manufacturer for repair without being charged the price of a new scope. If you want to try this, just ask your local Wal-Mart or K-Mart who they sell their return merchandise to, and they should be able to direct you right to their door.

4. Military surplus arms dealers, such as Navy Arms Corp. Purchase a military surplus scope from one of these stores. Unfortunately, you will never find an American military scope this way. The demand is so high that any scopes released by the government would be snapped up immediately. However, military scopes designed primarily for Russian and Chinese SKS rifles have been coming on the market recently. They appear to be good scopes, but they have a couple of problems associated with them. First they are around 2.5x, magnifications limited to around 200-250 yards. Navy Arms carries these (689 Bergen Blvd., Ridgefield, New Jersey); they are definitely short range scopes. The second problem is that they have 3/4 inch diameter tubes, which means that Russian or Chinese military mounts and rings will have to be used. This effectively limits their use (for the most part) to SKS rifles. It would be possible to shim a standard set of 1 inch rings to hold the 3/4 inch tube of the Chinese scope, but it would require having a machine shop machine a 1/4 inch thick shim with an outside curvature diameter of 1 inch and an inside curvature diameter of 3/4 inch. Whether this would be worth the cost is something that the shooter would have to decide for himself. If this is a way that you might want to go, it would be a good idea to supply the machine shop with a set of 1 inch rings and the Chinese or Russian scope, so they can see precisely what you are trying to do.

5. Go through a FFL licensed gunsmith or firearms dealer, and negotiate the cost + 10% that was described earlier in the manual for rifle purchases. This is the best alternative, because it allows you, the consumer, to settle on exactly what you want at exactly the price you want. You control everything including the brand and quality of the scope, and you CAN get a very good scope for a very good price.

Scope Mounts

Once you get your scope you will need to purchase a mount. You may spend nearly as much on the mounts as you will on the scope itself. Although mounts are available through the sporting good sections of retail chains such as Wal-Mart, K-Mart or Target, these are usually scope bases (minimal mounts) which require drilling and tapping of the rifle, and in the case of certain weapons (such as the pattern 17 Enfield), the rear of the receiver may have to be machined down to a certain configuration. The bases may be relatively inexpensive, but you usually pay for the convenience in other ways. Commercial rifles are often drilled and tapped for scope mounts. If your rifle fits this category, get the cheaper bases. If it is NOT already drilled and tapped for scope bases, I recommend using military conversion mounts made by such companies as B-Square or S&K (available through Brownells). These mounts are designed to take advantage of the stock military configuration of various rifles, and they attach without any drilling or tapping. This is advantageous for several reasons. First of all, some rifles have incredibly hard receivers. Springfield rifles are a good example. Receivers on these rifles are so hard that drilling alone is a major job, and may require heat treating prior to drilling just to get the metal soft enough to drill. Drilling and tapping is not necessarily as simple as it seems on the surface. Receivers such as this are a headache for a gunsmith, so you can imagine what it would be like for someone else. Also, many military rifles have collector value, and once they are drilled and tapped for scope bases, the collector value is gone. For these reasons (and others as well) I recommend the use of "no drilling or tapping" scope mounts when available. Mounts of this type (known under the S&K label as Insta-Mounts) are available for the 30-40 Krag, the SMLEs, the pattern 14 & 17 Enfields and the Springfield rifles (Springfields require minor grinding for clearance).

Mauser rifles present a problem when it comes to scope mounting. First of all, some Mausers come with a bolt handle which protrudes straight out from the side of the rifle. These bolt handles will hit the scope when the action is opened. The bolt handle problem is solved by forging the handle to a bent-down configuration. Unfortunately this is something that a gunsmith will have to do for you.

The S & K Insta-Mount readily fits many old military rifles without the need for drilling the receiver. Photo courtesy S & K.

Also, when regular riflescopes are mounted on a Mauser, there is insufficient clearance to work the safety. The safety problem can be resolved in any one of four ways. One solution would be through the installation of a Timney Featherweight Deluxe trigger group (See discussion on triggers, covered earlier). This puts the safety on the trigger instead of on the rear of the cocking piece. Another solution would be to install a side swing safety, but this generally requires machining of certain parts, gunsmith installation, and it is expensive. Parts will cost from $30 to about $150, and this doesn't include the labor of the gunsmith. A better solution is the use of a Buehler low-safety (available for the Springfield, the Krag and most Mausers), which can be purchased for less than $30, and installed in the bolt in the place of the old safety. This system engages the safety with a small arc movement (70 degrees), making the normal 180 degree or so movement unnecessary. The fourth solution would be the installation of a B-Square no-drill mount. This mount replaces the rear leaf sight, and requires a long eye relief (LER) pistol scope. These scopes do not have the magnification necessary for long distance shots, as the highest magnification for LER scopes is about 3x, with 1.5x to 2.5x being more common. Also, the weight of the scope being placed that far forward creates a balance problem, but I must confess, this may be a personal idiosyncracy unique to myself. Jeff Cooper likes the forward mount and finds balance (for him) improved over normal scope placement. Of course, the good thing about the B-Square no-drill mount is that you don't have to drill or tap any holes, you don't have to alter bolt handles and you don't have to mess with the safeties. Parts for all four solutions are available from Brownells.

Using the B-Square no-drill mount on the Mauser requires a long eye relief (LER) pistol scope such as this Bausch & Lomb Elite 3000 handgun scope. These scopes do not have the magnification necessary for long distance shots, as the highest magnification for LER scopes is about 3x, with 1.5x to 2.5x being more common. Photo courtesy of Bausch & Lomb Sports Optics Division.

Moisin Nagants, either Russian or Finnish, present a unique problem when it comes to scope mounting. This rifle has a split receiver. By this, I mean that the slot in which the bolt handle travels runs the full length of the back 3/4 of the receiver, effectively separating it into two halves. The bolt handle is one that extends straight out to the side horizontally, and when the action is opened it rotates vertically into the receiver slot. There is no bridge over this slot, so there is no place on the top of the receiver which will allow the normal mounting of a telescopic sight. Actual Moisin Nagant sniper rifles used a side mount with a very high scope placement to allow for the normal operation of the bolt handle. Since these mounts are generally not available, and since there are no commercial mounts for the rifle, the best solution would be to make a scope base which replaces the rear sight leaf, and install a long eye relief scope on it. Making this scope base is very simple, using a file, a hacksaw, regular Weaver scope bases and a piece of aluminum barstock. Select the aluminum bar stock in a width that allows it to fit precisely between the sight ears through which the rear sight leaf is pinned. Cut the stock to the length of the rear leaf sight. File the ends of the barstock to remove hacksaw marks. Drill a horizontal hole crossways through the end of the bar to allow the pinning of the bar to the rifle barrel. Next drill a vertical hole through the back of the base to use the existing screw hole which holds the leaf sight mount to the barrel. Position the Weaver scope bases on the bar spaced an appropriate distance apart, center punch the holes in the bases on the aluminum bar, drill and tap the holes for the Weaver scope base screws (aluminum drills and taps very easily). Mount the scope bases on the bar, then spray the whole thing with black epoxy paint to finish it. This is a very simple process that anyone can do.

Scope Rings

Once you have the scope of your choice and the mount, you will need to obtain a set of rings. The rings hold the scope and provide an attachment point for the mount or scope base. Unfortunately, there are a large number of ring types, and few of these different types are compatible with one another. Weaver rings are the most common, but make sure you follow the instructions of the mount manufacturer in this regard. If the manufacturer says to use Weaver rings, don't purchase Conetrol rings, because they won't work in the mount. Fortunately, scope rings are incredibly common, and are available from many hardware stores and sporting goods stores, as well as gunsmith suppliers. Even Wal-Mart and K-Mart will have generic brands of scope rings which are designed to fit Weaver bases.

Also, scope tube diameters differ, making different ring diameters necessary. Make sure that you get the right diameter for your scope. A one inch tube is the most common, but larger diameters have been coming out in recent years.

For sniper conversions one type of ring should be avoided. This is the high mount see-through ring used by hunters that allows the shooter to either use the scope by positioning his head very high, or to look through an opening between the scope and the receiver top to use the open iron sights in the event of a scope failure. Although this sounds good in theory, it does not work out in a sniper situation. For a sniper use, the scope needs to be mounted as close to the top of the receiver as possible. This type of mount will maximize both accuracy and comfort to the sniper, if he has to spend long hours peering through his scope.

Scoping the Number 4 SMLE Test Rifle

It was decided that some of the more expensive (and better) scopes would be avoided as a choice for our test rifle, so that we could try to maximize function and effectiveness while minimizing cost. With this in mind, the scope chosen for our test rifle was a Bushnell Buckhorn 3X9 variable scope, with a 32 mm. diameter objective (a 40 mm objective would have been a better choice, as larger objectives gather more light for twilight shots). The scope was purchased new from Nationwide Sports for $35.90 wholesale. This scope boasts of a wide angle field of view, 1/4 MOA click adjustments, fully coated optics, a one piece body tube, waterproof and fogproof, and fully warranted for the life of the original owner. Notably, it did not possess external adjustment knobs. The scope mount chosen was an S&K Insta-Mount for the number 4 SMLE, purchased at $28 wholesale. I had a set of Weaver rings from a scoped rifle I owned many years ago, so rings (in this case) cost me nothing. The price of a military surplus rifle with a bore "like new" plus the cost of the scope plus the cost of the scope mount came to a total cost of $125.80. The total cost of materials used in glass bedding and trigger adjustment was less than $10. Of course, we are not considering labor costs because this is something the reader will provide himself. Add to these totals 10% and the cost of a set of rings, and the reader has approximately what the cost would be to him.

Installation of the Insta-Mount was simple:

1. The SMLE's bolt, the rear sight and the spring loaded plunger found under the rear sight are removed.

2. The screws and the wedge shaped locking nut on the bottom of the Insta-Mount are removed.

3. The mount base is positioned on the receiver, and the two screws turned loosely into the Insta-Mount through the peepsight mount holes in the rear of the receiver.

4. The scope mount is pivoted downward, while the locking nut (with the angled bevel positioned toward the rear of the receiver) is slipped into place.

5. The cap screw which turns into the locking nut is tightened, as well as the two screws in the rear of the Insta-Mount. Blue loctite can be placed on the threads prior to tightening to ensure the screws do not shoot loose.

6. The bottom half of the Weaver rings were positioned in the appropriate horizontal slots cut into the Insta-Mount, and tightened into place.

7. The scope was placed on the two ring halves with the adjustment apertures centered between the rings. The top half of the rings was positioned and ring machine screws inserted and started in their threads. Looking through the scope, the tube is turned until the crosshairs run vertically and horizontally, then the rings are tightened down evenly.

This completes the installation of the scope. Adjustments for windage and elevation are made at the top and right side adjustment sites on the scope using a coin or a small screwdriver, if your scope doesn't come equipped with adjustment knobs.

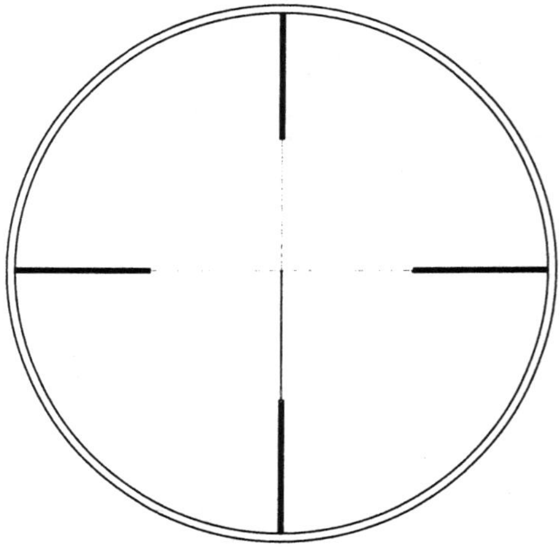

CHAPTER 6

The Final Touches

A few additional comments are necessary to round out the sniper rifle. First is the importance of rangefinding. If you anticipate the need for a two man team as is currently used by the military, one will need a rangefinder. Excellent laser rangefinding binoculars are on the market at the present time, but they will cost upward of a thousand dollars. A rangefinder calibrated out to 1000 yards is made for hunting and is available for around $100. Check with large sporting good stores for this item. When coupled with an ordinary set of binoculars, this combo can function quite well at a fraction of the cost.

If you are anticipating being alone in your sniping position, it is a good idea to select a spot with a good back door for a controlled retreat, then pace off the distance to certain landmarks so that you always have a fairly good idea of how close targets are to your position.

If your scope does not contain external knobs for

The LaseRange SR is advertised to have an effective range of 200 yards but it has been reported that the distance can extend up to 600 yards on highly reflective surfaces. Photo courtesy Ranging, East Bloomfield, NY.

The 1000 Rangematic rangerfinder as used by hunters is an example of a quality rangerfinder that is affordable and offers effectiveness out to 1000 yards. Photo courtesy Ranging, Brookfield, NY

adjusting windage and elevation, the best way to use it is to zero the scope at an optimum range and leave it at this setting, making adjustments for reasonable distances from the knowledge of bullet drop trajectories. For example, in the case of our test SMLE, I would recommend that the shooter zero the rifle at 200 yards. This will put his shots about 5 inches high at 100 yards., right on at 200 yards, about 7 inches low at 300 yards, a little over a foot low at 400 yards and slightly over two feet low at 500 yards. If the sniper knows the distances he is shooting at, he should be able to compensate for bullet drop out to 500 yards without much trouble. With the crosshairs held dead on to a point just below the hollow of the throat, all shots should be in the kill zone. For a 500 yard shot, the sniper would have to hold the crosshairs on the top of the head for a standing target. Unless the shooter is equipped with something like the Shepherd scope, he cannot realistically expect to be too accurate beyond the 500 yard mark.

If the shooter is equipped with a 30.06, the recommendations would differ. If I had a 30.06 sniping rifle, I would handload cartridges to about 2800 fps and use a Hornaday 168 grain boattail hollow point bullet. With this combination I would zero the rifle at 300 yards, and the resulting trajectory should result in bullets arriving about 1 ft. high at 100 yards, about 8 inches high at 200 yards, right on at 300 yards, about 8.5 inches low at 400 yards and about a foot and a half low at 500 yards. In this case you could probably hold the crosshairs to a point just below the base of the throat at all distances out to 500 yards. The main point here is that the shooter will have to look at the trajectory of his own cartridge and chosen bullet design and weight to determine the best place for zeroing his scope, i.e. the point that will make him the most efficient at both short and long ranges.

Another question that is often asked is that of whether a recoil pad should be put on the rifle. This will depend largely on the rifle and the person shooting it. If the rifle is a hard kicker, it may be advantageous to install one. All SMLE number 5 jungle rifles come factory equipped with recoil pads because their shortened size and resulting weight reductions create a particularly hard kicking rifle. It really depends on both the rifle AND the shooter. I would recommend NOT putting one on unless the recoil is truly uncomfortable. If you decide that it is necessary, put on a Pachmayr slip-on recoil pad and see if the rifle is comfortable to shoot. Slip-on pads generally don't last very long, but they are inexpensive and useful in determining whether adding a recoil pad will make the stock too long for the shooter. If it does, when you are ready to install a real recoil pad, you will have to cut the stock off to the recoil pad depth. Since some stocks are too short for many people to begin with (like the Chinese SKS and the Russian type 44 bolt action rifle), a recoil pad installed without shortening the stock could fit the bill exactly, essentially elongating the stock to a more comfortable fit for the shooter. The slip-on pad is a good way to find out without cutting the stock prematurely.

Another accoutrement which is recommended, but not necessarily a requirement, is a lace-on cheek piece. This is a padded leather affair that laces onto the stock just like a pair of shoes lace up. It positions the head of the shooter slightly higher on the stock so that his shooting eye falls naturally in line with the scope rather than at a lower position in line with where the iron sights would normally be. Such an item has been a standard accessory for military snipers over the years. Cabela's (812 13th Ave., Sidney, Nebraska, 69160) offers a very nice padded lace-on leather cheek pad for around $20. This item has the added advantage of being removable, so it can be readily transferred from rifle to rifle.

This manual will produce a highly accurate rifle at a nominal cost. I would bet my life on the accuracy and dependability of a rifle built like this. However, there is more to sniping than a highly accurate rifle. It is recommended that additional texts covering other aspects of sniping be consulted, such as : **U.S.M.C. Sniping, Camouflage, and Guerrilla Warfare and Special Forces Operations , all available from Delta Press, Ltd., P.O. Box 1625, El Dorado, AR 71731-1625.**

If the reader has any questions about the material presented here, he may write to me in care of the publisher. Please include a self-addressed stamped envelope. I will answer all inquiries as time permits.

Dan Boone

Appendix

Gun Parts

Exploded Diagrams of Surplus Rifles

The following pages contain exploded diagrams and listings of the parts of each individual rifle. The listings of parts also contain prices which are not to be considered current. For current prices on any parts shown on the diagrams please contact Gun Parts Corporation who generously allowed the reproduction of the diagrams from their catalog.

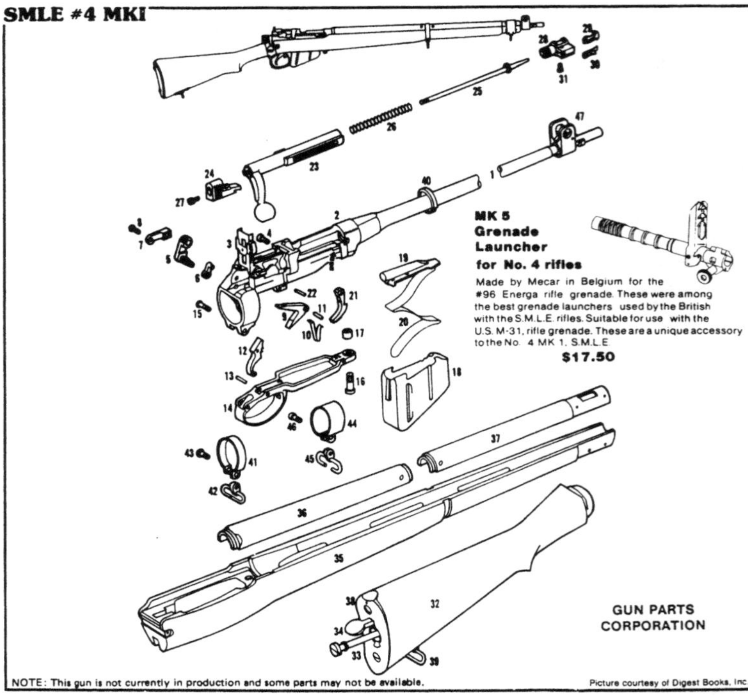

*Starred items not always available.

		List Price	Part No.	Description	List Price	Part No.	Description	List Price
1	Barrel - used	$39.95	19	Follower	$5.50	39	Butt sling swivel assy	$4.05
2	Receiver FFL Required	*	20	Follower spring	2.30	40	Rear stock band	3.20
3	Rear Sight — Milled	11.25	21	Magazine catch	2.90	41	Middle stock band	3.75
	Stamped	9.65	22	Magazine catch pin	1.10	42	Front sling swivel	1.25
	"L" Flip	8.20	23A	Bolt stripped	17.20	43	Front sling swivel screw	1.25
4	Ejector screw	1.50	23B	Bolt Complete	37.25	44	Front stock band	3.60
5	Safety bolt	4.75	24	Cocking piece	6.95	45	Stacking swivel	1.65
6	Safety catch	4.95	25	Firing pin	10.00	46	Stacking swivel screw	1.10
7	Safety bolt spring	2.60	26	Firing pin spring	2.25	47	Front sight guard	3.75
8	Safety bolt spring screw	4.05	27	Firing pin stop screw	2.25	48	Front sight base	3.25
9	Sear	3.80	28	Bolt head stripped	8.50	49	Front sight base & guard screw	1.15
10	Sear spring	4.25	29	Extractor	6.95	50	No. 4 MKI bolt head release	4.05
11	Sear pin	1.10	30	Extractor spring	4.25	51	No. 4 MKI bolt head release spring	1.75
12	Trigger	3.20	31	Extractor screw	2.25	52	Bolt head release stop	2.25
13	Trigger pin	1.10	32	Stock butt	22.95	53	Bolt head release stop screw	1.75
14	Trigger guard	6.35	33	Stock bolt	1.50	54	Rear sight base pin	1.35
15	Rear guard screw	1.50	34	Stock bolt washer	1.00	55	Front sight blades	1.30
16	Front guard screw	1.50	35	Forestock	26.95	56	Rear sight plunger	1.75
17	Front guard screw bushing	1.00	36	Rear handguard	5.50	57	Rear sight plunger spring	1.15
18	Magazine complete	18.00	37	Front handguard	5.50	58	Stripper Clips	1.00
			38	Buttplate	7.50			

375

SMLE #4 MK1

Diagram compliments of Gun Parts Corporation, West Hurley, N. Y. 12491 Note: Prices may not be current.

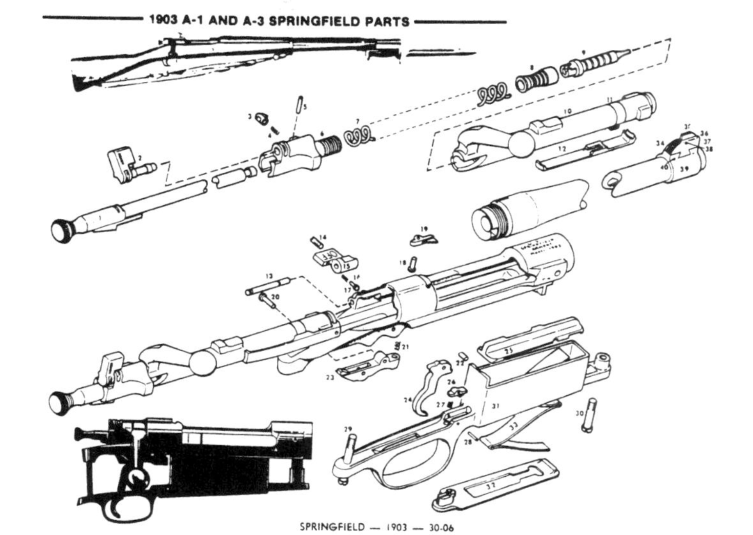

GUN PARTS CORPORATION

*Quantities limited.

Part No.	Description	Price
SP1	Firing Pin Rod	$4.75
SP2	Safety Lock Assy	2.60
SP3	Bolt Sleeve Lock	1.75
SP4	Bolt Sleeve Lock Spring	.95
SP5	Bolt Sleeve Lock Pin	.95
SP6	Bolt Sleeve (Stripped)	4.35
SP6A	Bolt Sleeve (Complete)	6.85
SP7	Mainspring	3.30
SP8	Striker Sleeve	2.30
SP9	Striker	4.75
SP11	Extractor Collar	3.60
SP12	Extractor	8.75
SP13	Cut-off Spindle	2.60
SP14	Cut-off Screw	1.15
SP15	Cut-off	2.90
SP16	Cut-off Plunger Spring	.95
SP17	Cut-off Plunger	.95
SP18	Ejector Pin	1.40
SP19	Ejector	3.50
SP20	Sear Pin	1.45
SP21	Sear Spring	1.15
SP22	Trigger Pin	1.15
SP23	Sear	4.20
SP24	Trigger	4.20
SP25	Fóllower Milled	4.55
SP26	Floor Plate Catch	2.05
SP27	Floot Plate Catch Spring	.95
SP28	Floor Plate Catch Pin	.95
SP29	Rear Guard Screw	1.15
SP29A	Rear Guard Screw Bushing	1.15
SP30	Front Guard Screw	1.15
SP31	Trigger Guard Milled	19.50
SP32	Floor Plate	4.55

Part No.	Description	Price
SP33	Magazine Spring	$1.75
SP34	Front Sight Base	3.90
SP35	Front Sight Blade	1.15
SP36	Front Sight Base Screw	1.15
SP37	Front Sight Base Spline	1.15
SP38	Front Sight Blade Pin	.95
SP39	Front Sight Band	3.20
SP40	Front Sight Base Pin	1.05
SP40A	Front Sight Assy Complete	7.50
SP41	Front Sight Cover	4.60
SP41A	USMC Front Sight Cover	8.95
SP42	Rear Sight Leafs	2.90
SP43	Rear Sight Leaf Spring	1.75
SP44	Rear Sight Binding Knob, Ele. Slide	1.75
SP45	Rear Sight Windage Knob Assy	2.30
SP46	Rear Sight Leaf Base	3.95
SP47	Rear Sight Complete Less Base Collar	7.50
SP48	Rear Sight Base Collar	9.25
SP49	Rear Sight Drift	2.30
SP50	Butt Plate, Checkered Milled	12.10
SP50A	Butt Plate, Smooth (Old Style)	8.05
SP51	Butt Plate Trap Door, Checkered	3.40
SP51A	Butt Plate Trap Door, Smooth	2.90
SP52	Butt Plate Trap Door, Springs	2.30
SP53	Butt Plate Screws, Small (also fit butt swivel)	1.15
SP54	Butt Plate Screws, Large	1.75
SP55	Stacking Swivel Screws	1.15

Part No.	Description	Price
SP56	Sling Swivel Screws	$1.15
SP57	Sling Swivel	1.45
SP58	Butt Swivel Milled	2.90
SP59	Upper Band Screws	1.15
SP60	Stacking Swivel	1.45
SP61	Stock Reinforcing Nu	1.90
SP62	Stock Reinforcing Bolt	1.90
SP63	Lower Band Spring	1.75
SP64	Upper Band	7.50
SP65	Lower Band	7.50
SP66	Bolt Stop (Early 03 only)	4.50
SP67	Hand Guard	4.05
SP67A	Hand Guard Clips	.95

Barrels & Stocks

Part No.	Description	Price
SP 150	Barrel 03A3 4-Groove	59.50
SP 155	Barrel 03A3 new 2-Groove	44.95
SP 161	Stock Straight Grip	32.20
SP 162	Stock Straight Grip (Refinished)	37.95
SP 10	Bolt Stripped, 03A3	9.5
SP 10A	Bolt Stripped, 03	12.9
SP 10B	Bolt Stripped, 03, Nickel Steel	15.9

All models interchange.

Diagram compliments of Gun Parts Corporation, West Hurley, N. Y. 12491 Note: Prices may not be current.

Part No.	Part Name	Price
SP69	Trigger Guards, 1-Piece, Stamped	$9.95
SP70	Butt Plates	5.20
SP71	Upper Band	6.95
SP72	Lower Band	4.50
SP73	Follower	3.75

Part No.	Part Name	Price
SP74A	Rear Sight Base Screw	$1.15
SP74	Rear Sight	$6.65
SP75	Front Sight Key	1.15
SP76	Upper Band Screw	1.45
SP77	Front Sight Blade	1.15
SP78	Front Sight Base	2.90
SP79	Hand Guard Ring	4.25

Part No.	Part Name	Price
SP80	Butt Swivel	$2.60
SP81	Stacking Swivel (Stamped)	1.45
SP82	Stacking Swivel Band	5.95
SP83	Sling Swivel (Stamped)	1.45
SP84	Hand Guards, Wood	4.05

Part No.	Part Name	Price
SP85	5 Shot Stripper Clips M1903	4/$1.00
SP86	Ruptured Cartridge Extractor	4.55
SP87	Oil Bottles, nickel plated brass	6.50
SP88	Thongs	95
SP89	Take Down Tool	3.00
SP90	Hand Guards, Composition	5.30
SP91	Stocks, Compositions	48.50
SP92	Safety Assy. Tool	5.70
SP93	Field Test Bolt Tool	10.30
SP95	Dry Firing Practice Piece	1.75

Accessories

Part No.	Part Name	Price
SP99	Low Scope Safety	4.99
SP107	Coaching Device 03—as issued	10.30

GUN PARTS CORPORATION

Part No.	Part Name	Price
SP108	Bent Bolt Stripped, Blued	$14.95
SP109	Bent Bolt Complete, Blued	29.75
SP110	Military Style Leather Sling (replacement)	12.95
SP122	Redfield Base Windage Screw	.95
SP130	Inletting Guide Screws	2 for 1.50
SP72A	Lower Band Spring	1.75
SP133	Short Follower	4.95

Part No.	Part Name		Price
SP112	Grips, Plastic, left or right	ea.	2.75

Bayonet Parts

Part No.	Part Name	Price
SP115	Hinge Bar	3.75
SP117	Release Catch Spring	.95
SP118	M7 Scabbard for 10'' Blade	7.50

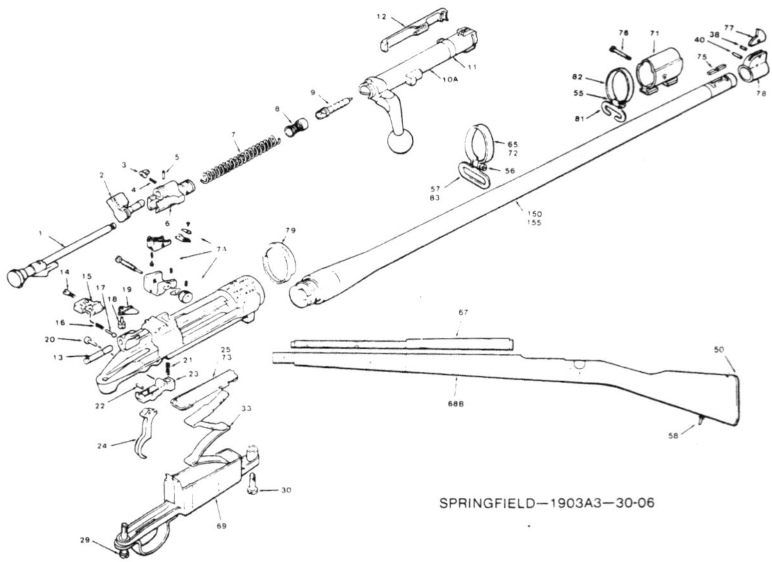

SPRINGFIELD—1903A3—30-06

378

Diagram compliments of Gun Parts Corporation, West Hurley, N. Y. 12491 Note: Prices may not be current.

U.S. ENFIELD 1917 30-06

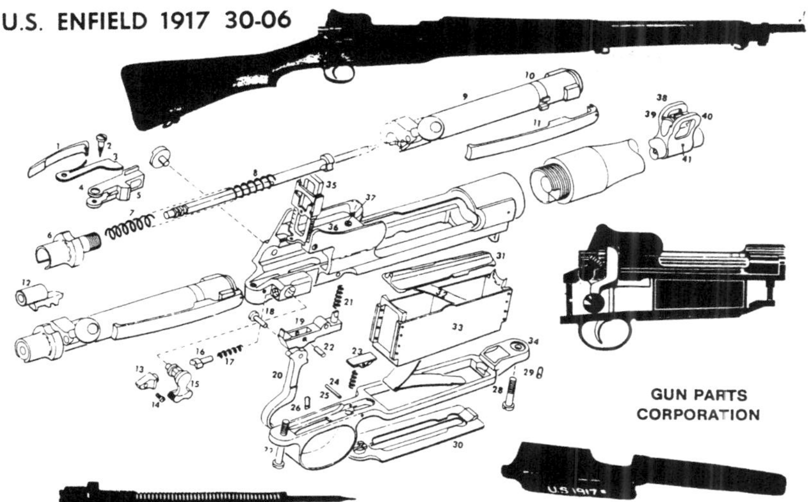

ENFIELD SPEED LOCK
Cocks-on-Opening

FOLLOWER DEPRESSOR FOR U. S. M1917 RIFLE
Keeps follower down so bolt rides over follower.
Rare Accessory. **$9.65**

(Receivers Marked "Remington", "Winchester" and "Eddystone")

Part No.	Description	Price	Part No.	Description	Price	Part No.	Description	Price
E17-1	Bolt Stop Spring	2.90	E17-24	Floorplate Catch Spring	1.05	E17-46	Butt Plate Screw (Small)	1.05
E17-2	Bolt Stop Screw	1.50	E17-26	Rear Guard Screw Bushing	1.30	E17-47	Rear Swivel Base	1.95
E17-3	Ejector	8.95	E17-27	Rear Guard Screw	1.95	E17-48	Rear Swivel Assy.	3.55
E17-4	Bolt Stop Spring Rest	2.55	E17-28	Front Guard Screw	1.95	E17-49	Rear Swivel Stock Screw	1.05
E17-5	Bolt Stop	4.15	E17-29	Front Guard Screw Bushing	1.05	E17-50	Handguard Ring	3.90
E17-6	Sleeve Bolt	6.70	E17-30	Floorplate	4.80	E17-51	Lower Band	3.25
E17-7	Mainspring	3.55	E17-31	Follower	5.05	E17-52	Lower Band Screw	1.05
E17-8	Firing Pin	12.05	E17-32	Magazine Spring	1.60	E17-53	Stacking Swivel	2.75
E17-9	Bolt Stripped	13.20	E17-33	Magazine Box	6.95	E17-54	Stacking Swivel Screw	1.05
E17-9A	Bolt Complete	34.95	E17-34	Trigger Guard	8.60	E17-55	Upper Band	3.75
E17-10	Extractor Collar	3.60	E17-35	Rear Sight Assy.	10.10	E17-56	Upper Band Screw	1.05
E17-11	Extractor	9.75	E17-35A	Rear Sight Slide Complete	4.95	E17-57	Follower Depressor	9.65
E17-12	Cocking Piece	4.30	E17-35B	Rear Sight Slide	3.20	E17-58	N.A.C. Cock-on-Opening Speed Lock (P-17)	16.45
E17-13	Safety Lock Holder	3.75	E17-35C	Rear Sight Slide Spring	1.05	E17-59	Stock, New Walnut	19.95
E17-14	Safety Lock Holder Screw	1.30	E17-36	Rear Sight Base Spring	2.55	E17-59A	Birch Stock, New	9.95
E17-15	Safety	4.80	E17-37	Rear Sight Base Screw	1.05	E17-60	Handguard Lower, New	4.95
E17-16	Safety Lock Plunger	1.60	E17-38	Rear Sight Axis Screw	1.05	E17-61	Handguard Upper, New	4.95
E17-17	Safety Lock Plunger Spring	1.05	E17-39	Rear Sight Axis Nut	1.05	E17-64	Oiler, Butt Trap, Complete	5.50
E17-18	Sear Pin	1.05	E17-40	Front Sight Carrier (Base)	3.80	E17-65	Barrel	*
E17-19	Sear	4.15	E17-41	Front Sight Blade	2.90			
E17-20	Trigger	3.10	E17-42	Front Sight Spline	1.95			
E17-21	Sear Spring	1.10	E17-43	Front Sight Pin	1.30			
E17-22	Trigger Pin	1.05	E17-44	Butt Plate w/Trapdoor Assy	4.80			
E17-23	Floorplate Catch	2.50	E17-44A	Plate Spring	1.30			
E17-25	Floorplate Catch Pin	1.05	E17-45	Butt Plate Screws (Large)	1.45			

Below Parts Are for The British P-14 .303 Rifles—All Others Interchange With The P-17 Parts Listed Above

Part No.	Description	Price	Part No.	Description	Price	Part No.	Description	Price
E14-66	Trigger Guard	8.90	E14-71	Bolt Rest for Long Range Sight	2.90	E14-74	Bolt (Stripped)	11.7
E14-67	Firing Pin	12.05	E14-72	Long Range Front Sight	3.80	E14-75	N.A.C. Cock on Opening Speed Lock	16.4
E14-68	Follower, Winchester	5.05	E14-73	Long Range Front Sight Screw	1.05	E14-76	Bolt Stop Spring for Long Range Sight.	2.9
E14-69	Extractor	24.75				E14-78	Ejector	14.95
E14-70	Box Magazine	5.40				E14-68B	Follower, Eddystone	5.05

*NOTE The gun is not currently in production and some parts may not be available
*Starred items not always available

382

Diagram compliments of Gun Parts Corporation, West Hurley, N. Y. 12491 Note: Prices may not be current.

SMLE #1 MKIII

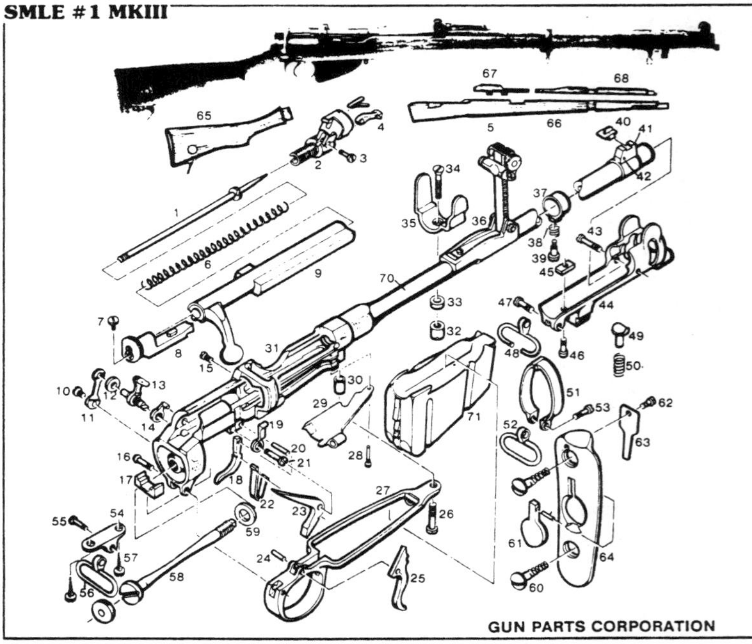

NOTE: This gun is not currently in production and some parts may not be available.

*Starred items not always available

Part No.	Description	List Price
1	Firing pin	$10.00
2	Bolt Head, stripped	10.00
3	Extractor screw	2.25
4	Extractor	6.95
5	Extractor spring	4.25
6	Firing pin spring	2.25
7	Firing pin screw	2.25
8	Cocking piece	6.95
9A	Bolt body stripped	17.20
9B	Bolt complete	37.95
10	Safety spring screw	1.75
11	Safety spring	2.60
12	Safety catch washer	1.50
13	Locking bolt (safety lever)	3.80
14	Safety catch	4.90
15	Ejector	2.00
16	Rear trigger guard screw	1.50
17	Stock bolt plate	2.90
18	Magazine catch	2.90
19	Bolt head retaining spring	3.10
20	Magazine catch pin	1.10
21	Retaining spring screw	1.50
22	Sear spring	4.25
23	Sear	3.80
24	Trigger pin	1.10
25	Trigger	3.20

Part No.	Description	List Price
26	Guard screw	$1.10
27	Trigger guard	7.50
28	Cut off screw	1.10
29	Cut off	7.50
30	Guard screw bushing	1.00
31	Receiver (FFL required)	22.95
32	Protector nut	1.50
33	Forend collar	1.50
34	Rear sight protector screw	1.75
35	Rear sight protector	2.90
36	Rear Sight Assembly	9.95
36A	Rear sight adj. elevation only	6.85
36B	Rear sight adj. windage, elev.	8.00
36C	Rear sight base	5.40
36D	Rear sight leaf	2.60
36E	Rear sight base screw	1.25
36F	Rear sight hingescrew and nut	1.70
37	Inner band	3.75
38	Inner band spring	1.00
39	Inner band screw	1.75
40	Front sight blade	1.75
41	Front sight base	2.60
42	Front sight pin	1.00
43	Front nose cap screw	1.75
44	Nose cap	5.50
45	Nose cap nut	1.75
46	Rear Nose cap screw	1.50

Part No.	Description	List Price
47	Stacking swivel screw	$1.10
48	Stacking swivel	1.65
49	Forend stud	1.75
50	Forend stud spring	1.10
51	Swivel band	2.90
52	Front swivel	1.30
53	Swivel screw	1.00
54	Butt swivel base	3.50
55	Swivel base screw	1.25
56	Buttstock swivel	2.25
57	Butt swivel screws	1.00
58	Stock bolt	1.95
59	Stock bolt washer	1.00
60	Buttplate screw	1.50
61	Buttplate trap door	1.50
62	Trap door screw	1.00
63	Trap door spring	1.00
64A	Buttplate (complete)	7.50
64B	Buttplate (stripped)	5.00
65	Buttstock	22.95
*66	Forestock	*
67	Rear handguard	9.95
68	Front handguard	11.45
69	Stripper clip	1.00
70	Barrel, used	*
71	Magazine	18.00

374

Diagram compliments of Gun Parts Corporation, West Hurley, N. Y. 12491 Note: Prices may not be current.

GERMAN MAUSER 98

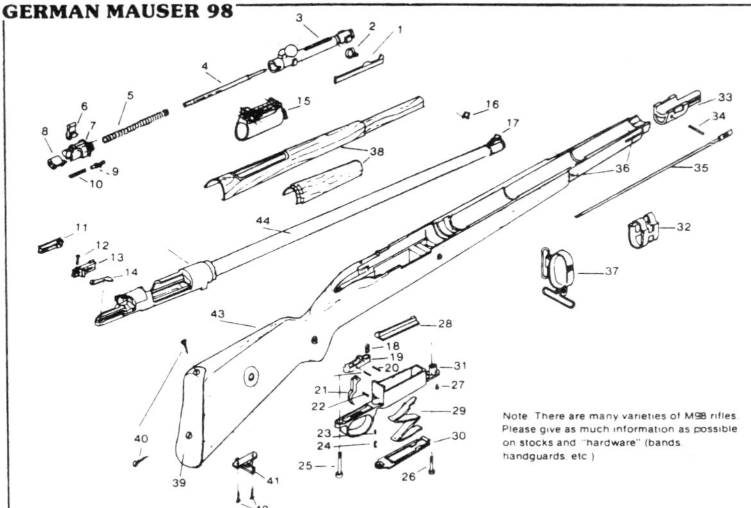

Note: There are many varieties of M98 rifles. Please give as much information as possible on stocks and "hardware" (bands, handguards, etc.)

Part No.	Description	List Price
1	Extractor	11.95
2	Extractor collar	3.25
3	Stripped bolts (military)	
	A - Straight handle	24.95
	B - Curved handle	24.95
	D - Low scope bent bolts mag cal	24.95
4A	Firing pin, military	8.95
4B	Firing pin, new	9.95
5	Firing pin spring	3.50
6A	Safety	3.20
6B	Low scope safety	4.99
	Bolt sleeve stripped	9.25
8	Cocking piece	4.95
9	Bolt sleeve catch	3.25
10	Bolt Sleeve catch spring	1.25
11A	Ejector spring and bolt stop cover assy	12.95
11B	Ejector spring	•
11C	Bolt stop cover	•
12	Bolt stop screw	1.15
13	Bolt stop	5.95
14	Ejector	4.50
15A	98K rear sight complete	7.50
15B	98K rear sight leaf	2.30
15C	98K rear sight slide	2.30
15D	98K rear sight base spring	1.25
15E	98K rear sight base (band)	2.60
15F	98K sight ramps	3.45 pr.
15G	98K rear sight screw	1.15
15H	GEW98 sight complete, early style	10.35
15O	98-09 Argentine rear sights (complete)	8.35
15P	98-09 Argentine rear sight screws	.95
16A	Front sight blade	2.95
16B	Front sight blade 98A	3.05
17A	Front sight complete	4.35
17B	Front sight base	2.30
17C	Front sight screw	1.15
17D	98A type front sight complete	4.90
17E	98A type front sight base	$3.45
17F	98A type front sight screw	1.15
17G	Muzzle cover	6.95
17H	Muzzle Cover 1904	4.75
18	Sear spring	1.10
19	Sear	3.20
20	Trigger pin	1.10
21	Trigger	3.20
22	Floorplate catch pin	1.10
23	Floorplate catch spring	1.10
24A	Floorplate catch	1.75
24B	External floorplate catch	3.75
25A	Rear guard screw military type	1.50
25B	Rear guard screw commercial	1.50
26A	Front guard screw military	1.50
26B	Front guard screw commercial	1.50
27	Lock screw for military only	1.25 ea.
28A	Follower-stamped	3.95
28B	Follower-milled	6.95
28C	Follower-short for use with magazine spacer in conversions	4.25
29A	Magazine spring	1.75
29B	Magazine spring for use w/short follower & mag. spacer	5.95
30A	Floor plate-stamped, late War issue	5.75
30B	Floor plate-milled	8.25
31A	Trigger guard milled military type less floor plate	19.95
31B	Trigger guard commercial w/mag. spacer less floor plate	19.95
31C	Trigger guard commercial w/hinged floorplate	•
31D	Trigger guard stamped w/Nazi proofs, floorplate, catch complete	22.95
31E	Trigger Guard, stamped winter	19.95
31F	Trigger Guards, Lengthened	24.50
	Stripper clips	$1.15
32A	G98 front band	4.05
32B	98K front band	4.35
32C	98K front band late War issue	3.20
32D	Band raised in rear for handguard	3.75
32F	KAR 98A (hinged front band)	11.45
33A	Bayonet stud long approx. 3½ overall	4.35
33B	Bayonet stud short approx. 2¾	4.35
33C	Bayonet stud extension for use with short bayonet stud	1.45
33D	Bayonet stud extension pin	.95
34	Bayonet stud pin	.95
35A	Cleaning rod G98 type	8.50
35B	Cleaning rod 15½	3.20
35C	Cleaning rod 14⅞	3.20
36A	GEW 98 front band spring	2.60
36B	GEW lower band spring	2.60
36C	98K front & lower band springs 3⅝	5.75
36D	98K front & lower band spring 4	4.35
37A	Lower band with side & bottom swivels	3.20
37B	Lower band as above takes retaining screw	3.20
37C	Lower band retaining screw	.95
37D	KAR 98K lower band	3.75
37E	Lower band w/swivel	3.75
37F	98-09 Argentine lower band	3.75
38A	KAR 98K short handguard	9.95
39A	Buttplate flat old style	3.75
39B	Butt Plate, stamped, cupped	4.75
40	Butt screw	1.15
41A	Rear swivel	2.05
41B	Rear swivel leather piece	•
42	Rear swivel screws	1.15
43A	Stock G98 long type used	39.75

Also see stocks elsewhere in catalog.

NOTE: This gun is not currently in production and some parts may not be available.

222

Diagram compliments of Gun Parts Corporation, West Hurley, N. Y. 12491 Note: Prices may not be current.

Notes

Notes

Notes

Notes

Notes

Notes